TL791
.A33

BOUND FOR THE STARS

Saul J. Adelman
Benjamin Adelman

A SPECTRUM BOOK

Prentice-Hall, Inc., Englewood Cliffs, N.J. 07632

Library of Congress Cataloging in Publication Data

Adelman, Saul J
  Bound for the stars.

  (A Spectrum Book)
  Includes index.
  1. Astronautics. 2. Interplanetary voyages.
  3. Space colonies.   I. Adelman, Benjamin, joint author.
  II. Title.
  TL791.A33   629.4   80-17990
  ISBN 0-13-080390-1
  ISBN 0-13-080382-0 (pbk.)

© 1981 by Prentice-Hall, Inc., Englewood Cliffs, N.J. 07632

All rights reserved. No part of this book
may be reproduced in any form or
by any means without permission in writing
from the publisher.

A SPECTRUM BOOK

Printed in the United States of America

10 9 8 7 6 5 4 3 2 1

Editorial/production supervision and interior design by Maria Carella.
Manufacturing buyer: Barbara A. Frick.

PRENTICE-HALL INTERNATIONAL INC., *London*
PRENTICE-HALL OF AUSTRALIA PTY., LIMITED, *Sydney*
PRENTICE-HALL OF CANADA, LTD., *Toronto*
PRENTICE-HALL OF INDIA PRIVATE, LIMITED, *New Delhi*
PRENTICE-HALL OF JAPAN, INC., *Tokyo*
PRENTICE-HALL OF SOUTHEAST ASIA PTE., LTD, *Singapore*
WHITEHALL BOOKS, LIMITED, *Wellington, New Zealand*

To Kitty and Carol

CONTENTS

Preface   *xiii*

1
Introduction   *1*

2
The Mountain that Hangs by a Hair   *6*

3
Survey of the Solar System   *26*

4
The Conquest of Zero *g*   *37*

5
Propulsion:
The Key to Solar System Travel   55

6
Colonizing Earth-Moon Space   75

7
Solar System Navigation   105

8
Solar System Travel   121

9
Exploring and Colonizing Mars   128

10
Exploring the Solar System   155

11
Survey of Our Galaxy   174

12
Propulsion:
The Key to Interstellar Flight   180

13
Interstellar Navigation *210*

14
Detection of Extrasolar Planets *229*

15
The First Starship *249*

16
Organizing the First Interstellar Expedition *261*

17
The First Interstellar Expedition *272*

18
A Program for Space *290*

19
The Space Enterprise *309*

List of Acronyms *323*

Index *325*

PREFACE

Bound for the Stars surveys the major problems of solar system and interstellar space exploration and utilization—what is being done and what can be done. The ideas of many scientists and engineers are examined. It is not our intention to decide who is "right" and who is "wrong," since there are many alternatives. Rather, we suggest a practical course of action for the space enterprise. Our approach differs sharply from the rosy, unrealistic portrayals of the coming marvels of the Space Age painted in most popular books on space.

The Space Shuttle is a critical step forward in space transportation that can lead to the full use of Near-Earth space. To convert its potential into reality, the American people must understand and support the space effort. The space enterprise is not an act of charity—it is an investment in America's future. Many space operations, such as weather observing, the Landsat satellites, and communication satellite systems, have immediate payoffs that already amount to billions of dollars a year. Other activities, such as the missions of the planetary spacecraft, have revolutionized our knowledge of the planets. Spinoffs such as miniaturization of electronic parts are used in pocket calculators, television sets, digital wristwatches, computers, stereo sets, and numerous other devices that enrich our daily lives. The funds invested in NASA have paid off handsomely, both in national wealth and in the advancement of science.

What has been accomplished so far is only a beginning. In order to use and explore space, we must vastly expand our scientific and technological capabilities. Future benefits, such as inexhaustible fusion power, energy independence, and the end of air and water pollution could revolutionize American society, raising it to a new level of prosperity and sophistication.

This book first considers the barriers blocking the mastery of space, including weightlessness, the problems of living in space for years at a time, and the high cost of space transportation. It tells how they can be overcome by research in space medicine and development of permanent, self-sustaining life support systems and efficient rocket engines. Once these obstacles are surmounted, the enticing prospect of the exploration and colonization of the solar system becomes practical. From that stage, flight to the nearer stars becomes a feasible goal for advancing space science and technology.

Space offers endless possibilities, since mankind can never hope to explore more than a small fraction of the universe. Whether we take advantage of these wonderful opportunities is up to all of us.

We wish to thank Mr. Harold B. Finger, Dr. Robert L. Forward, Dr. Thora W. Halstead, Dr. Rufus R. Hessberg, Dr. David L. Morgan, Mr. Dan C. Popma, Mr. Francis C. Schwenk, Dr. Sebastian von Hoerner, and Mr. Joseph A. Wynecoop for helpful information and comments. We also wish to express our appreciation for similar cooperation to M. Yves Demerliac, Secretary-General of Eurospace; Mr. Tadahico Inada, Washington Representative, National Space Development Agency of Japan; and Mr. Wilfred J. Mellors, Head, Washington Office, European Space Agency. Any errors in the text are, of course, the responsibility of the authors.

Original illustrations, not otherwise credited, were done by Linda Kohl-Orton.

SAUL J. ADELMAN
Charleston, South Carolina

BENJAMIN ADELMAN
Silver Spring, Maryland

1 INTRODUCTION

We stand at the opening of the Space Age with the universe to explore, if we wish. Historians and social scientists have all noticed mankind's interest in exploration. In any society, there is almost always someone who is curious about what is over the next hill or mountain or what is across the river or the ocean. Explanations for this phenomenon range from simple curiosity to military considerations and include economic, scientific, and political motives.

Popular belief has chosen the discovery of America as the threshold event of the modern era—and correctly so, for the difference between the modern and the ancient-medieval outlook is in their views of the Earth. Until the Renaissance, the known world was a small part of a vast sphere surrounded by the dark, hostile unknown, whereas now the Earth is small and explicable.

Renaissance men undertook to finish the exploration of the Earth with the vehicle that made it possible, a dependable ocean sailing ship. Evolving for 5000 years in an inland sea, it embodied many inventions including clinker-built hull, two or more masts, keel, sternpost rudder, bowsprit, deck, rigging, pulleys, multisail masts, and canvas for sails. When improved by the Portuguese shipwrights in the early 1400s, it was ready to leave the security of inland and coastal waters and face the ocean. The Portuguese caravel was no larger than the Roman sailing ship but was far more seaworthy and able to sail into the wind. Navigation had attained the level of portolan charts, compass, astrolabe, cross-staff, and elementary spherical astronomy, so a knowledgeable ship captain with a competent navigator had a reasonable chance of reaching his destination and returning to his home port.

Prince Henry of Portugal, better known as Henry the Navigator, directed a planned program of exploration. His first expedition reached the island of Porto Santo in 1418. In 1420 the King of Portugal appointed him Master of the wealthy Order of Christ. Prince Henry used its funds to finance his plan to discover a sea route down the west coast of Africa to India. Expedition after expedition was sent out. The Canary and Azores Islands were discovered and then colonized. In 1434 Cape Bogador was reached and in 1444, Cape Verde. By the time of his death in 1460 the Portuguese had penetrated as far as Sierre Leone. The exploration of the unknown regions of the Earth was well under way.

The Portuguese seamen had to overcome their superstitions and the fear of imaginary dangers as well as the real ones of storms, shoals, reefs, hostile natives, and disease. In the 1300s the myths of antiquity, enshrined in the writings of Pliny the Elder and Solinus, were heavily reinforced by the stories of Sir John de Mandeville, a highly accomplished liar. The more ludicrous the myth, the more firmly it was believed. To their surprise, as the seamen sailed on, the Green Sea of Gloom and the Sea of Darkness never appeared. They found that the Niger River did not boil and that the tropical sea did not become a thick syrup, and they never did find ants as big as mastiffs when they landed on the African coast.

Do the difficulties that face interplanetary and interstellar travel resemble those of the first Renaissance ocean voyages? Are some of the dangers believed to face starship astronauts imaginary? Can we achieve the level of science and technology that will enable men to voyage deep into the space ocean, or must we remain confined forever to our solar system? These are questions that we investigate in this book.

In a way, most people are still medieval in their outlook, for they consider their home area to be the center of the universe. Often the most trivial local news is treated as earthshaking while important national and international events are minimized or ignored. Now that we are in the Space Age, we must view the Earth as our home from which we can explore the Solar System and then our Galaxy, the Milky Way system.

On July 20, 1969, two Americans, Neil Armstrong and Buzz Aldrin, landed on the Moon and so completed the first human voyage to another member of the Solar System. This achievement was a great technological accomplishment. Yet, measured by interstellar distances and the scientific and engineering problems to be solved before such voyages can be undertaken, it was a short and easy trip. Nevertheless, it marked the end of man's confinement to the Earth and its immediate environs.

Until recently, the finite extent of the Earth's surface area was not apparent even to educated people. There was always something new to be found in one of the Earth's unexplored regions. But the closing of earthly frontiers and the development of the airplane and the artificial satellite has ended the illusion of the infinite Earth. The first V-2s fired from White Sands, New Mexico, soon after the end of World War II carried cameras to altitudes of over 120 kilometers (75 miles). Their photographs showed the Earth's curvature. Later, the Mercury and Gemini astronauts reached 160 to 190 kilometers (110 to 120 miles), and their photographs and descriptions left no doubt that the Earth was a sphere with finite surface area. When the Apollo astronauts went to the Moon, we saw how small the Earth appears from our closest natural neighbor in space.

Throughout history, intelligent men have been interested in the Sun, the Moon, the planets, the comets, and the stars. With the exploration of the Earth completed, at least on a major scale, space has become for us what the western frontier was to our great-grandparents. The exploration of space promises for our own and future generations the excitement, vitality, and drive for technical advance that the wilderness frontier provided for Americans for 275 years.

One of the key questions of space travel is: "What is space?" Many people will say that it is the region of the universe beyond the Earth's atmosphere and between it and celestial objects such as the Moon, the Sun, the planets, and the stars. That is, space lies outside of the Earth. But this is a provincial viewpoint. If there are intelligent beings on another planet somewhere in the universe, they would conclude that the Earth is in space.

From the viewpoint of space exploration, space is that part of our universe that humans have, or will have, the ability to explore. Depending on the technology and discipline involved, there are various ways of naming regions of space. When the Space Age began in October 1957, the initial objective was to place a small satellite in Earth orbit a few hundred kilometers up. Later on, probes and then humans reached the Moon. Dr. Krafft A. Ehricke suggested that the region up to and including the orbit of the Moon should be called cislunar space. We call it Earth-Moon space. Beyond lies the realm of the inner solar system, which stretches from the Sun to beyond the orbit of Mars and the asteroid belt, and the outer solar system from the asteroid belt to beyond the orbit of Pluto, 40 times the distance of the Earth from the Sun.

The solar neighborhood or near-stellar space extends from beyond the Solar System to about 100 light-years from the Sun. This is the region for which large telescopes in space will be able to resolve the disks of stars and in which the first interstellar voyages will occur. Next comes our Galaxy, the Milky Way system, as a whole. Beyond are the Milky Way's satellite galaxies, which include the Large and the Small Magellanic Clouds, then the other galaxies of the Local Group, those of the Virgo Supercluster and the rest of the observable universe.

We occupy a middle ground in the universe in many respects. We are minuscule on a galactic scale and enormous on an atomic scale. From our human vantage point, scientists have been able to discover natural laws that govern matter from subatomic dimensions up to those of the universe as a whole. We are the logical products of the laws of nature. Minor changes in some of the fundamental physical constants would cause life as we know it to cease to exist.

We are reasonably confident that our science can describe—or at least in principle be extended to describe—adequately well how matter acts over this range of dimensions under almost all conditions that are expected to occur. The exceptions are for the most extreme conditions, such as ultracondensed matter in the centers of black holes, and for corrections to existing laws that will be manifest over distances comparable to the size of the universe. The laws of nature, as we know them, must also be at least reasonable approximations to the true or exact laws of nature in the realm in which we know them to operate. However, these gaps in our knowledge are not those anticipated to occur under the conditions of interstellar travel, even at a substantial fraction of the speed of light. By no means have all of the scientific and engineering problems for such ventures been solved, but logical extensions of our current science and engineering should allow us to overcome them if we wish to do so. In later chapters we will discuss some of these problems and what will be required to solve them.

Space exploration rests on science and technology. However, since few scientists have given it much consideration, our visualization of what space exploration will be like has come mainly from science fiction, such as the television series, *Star Trek*. Unfortunately, many science fiction writers violate the laws of nature in their stories, a practice they claim makes their stories more interesting. But this transforms their work from fiction to fantasy. Without a grounding in science and technology, it is difficult for a writer to visualize the possibilities realistically and to select those with dramatic potential. Because the public is poorly educated in science, anything even remotely plausible can be sold. Some writers, however, such as Verne, Wells, Hoyle, and Clarke, have shown that well-written stories consistent with contemporary science and technology can be highly popular.

Moreover, space exploration is transforming astronomy from a purely observational science to a laboratory and field science. Space exploration is the story of men traveling outward from the Earth to visit our celestial neighbors. Its future exploits will read like science fiction. Dr. Edward Teller has said that science is more interesting than science fiction. We agree. What could be more exciting than exploring a planetary system revolving around a distant star, receiving signals from another civilization, or finding life on another planet?

2
THE MOUNTAIN THAT HANGS BY A HAIR

For my part, I consider the earth very noble and admirable precisely because of the diverse alterations, changes, generations, etc., that occur in its incessantly. If, not being subject to any change, it were a vast desert of sand or a mountain of jasper, or if at the time of the flood the waters which covered it had frozen, and it had remained an enormous globe of ice where nothing was ever born or ever altered or changed, I should deem it a useless lump in the universe, devoid of activity and, in a word, superfluous and essentially nonexistent. This is exactly the difference between a living animal and a dead one; and I say the same of the moon, of Jupiter, and of all other world globes.

GALILEO[1]

The search for life outside the Earth has been a major incentive of space exploration. To some, it is the only one. However, this search has been so overplayed by the media that many people believe space exploration will be pointless if life is not found elsewhere in the solar system. This is a very dangerous assumption, for the absence of life

---

The chapter title is from the Talmud, Tractate Chagigah, page 10A.

from all other planets except the Earth would be very interesting, deeply significant for the theory of evolution, and an encouragement to human exploration and colonization of the solar system.

What Is Life?

What is life, the goal of this compelling search? Research in microbiology and the origin of life has narrowed the gap between living and nonliving. The viruses are a case in point. The tobacco virus can be stored like an inert chemical, yet, when applied to the leaves of a tobacco plant, cause a virus disease. For our purpose, life is defined as a self-reproducing system of giant molecules that carry on such functions as growth and metabolism (orderly chemical reactions) and that can evolve in time.

Suppose that there were another planet in the solar system like the Earth and not much closer to or further from the Sun. If this companion Earth were explored and found to be lifeless, scientists would be surprised and disturbed, while the life sciences would be shaken to their foundations. Or, if this para-Earth harbored exotic forms such as silicon-based organisms, would they not also be upset?

Life as we know it is based upon the chemistry of carbon. Carbon is unique among the elements in its ability to link its own atoms together in chains, rings, hexagons, helices, and other patterns with innumerable variations. Carbon compounds outnumber the compounds of all the other elements put together. More than two million carbon compounds have been discovered or synthesized, while over a trillion are possible. Only carbon can construct the wide range of substances needed to carry on the many functions of life. Carbon, moreover, does not form huge insoluble crystals like those formed by silicon, such as silicon dioxide (quartz and sand). The thousands of water-soluble carbon compounds can react quickly with one another and with other chemicals in solution. Water is the nearest substance to a universal solvent. That Earth life is based upon water-soluble carbon compounds is just what the laws of physics and chemistry would lead us to expect. Further, all living things on Earth have the same basic chemistry. The lifelessness of the other planets and moons, as found to date, supports the belief that life depends upon carbon and liquid water. From these considerations, we venture to make two predictions which are consistent with our present knowledge of the solar system:

1. Any planet, anywhere, that is found to have a liquid water ocean that is more than 100 million years old will contain life.
2. All planets without such oceans are lifeless.

Mercury, the nearest planet to the Sun, is a hot, airless waste that reaches over 500°C (900°F) during the day. No one seriously regards Mercury as a life-bearing planet.

The Mariner and Venera spacecraft discovered that the dense atmosphere of Venus exerts a pressure on its surface that is 90 times that of the Earth's atmosphere. The temperature at the surface is 450°C (840°F) and the atmosphere is 90 percent carbon dioxide. Drifting high overhead are clouds of sulfuric acid vapor. When robot vehicles, or men in tanklike vehicles, roam the terrain of Venus, they will not find life there.

Mars has been the popular favorite as a counter-Earth since the 1870s, when the distinguished astronomer, Giovanni V. Schiaparelli, discovered what he believed to be channels on Mars. In the early 1900s Percival Lowell concluded that Mars was criss-crossed by a network of canals constructed by beings with a human level of intelligence. The Mariner spacecraft, which returned the first detailed pictures of Mars, showed that the canals did not exist. The Viking landers, which were designed to discover whether life exists on Mars, revealed unusual chemical activity in the Martian soil but left the question of the presence of life there unsettled.

The upper atmosphere of Jupiter may contain a thin mist of prebiotic chemicals. Jupiter generates 2.5 times as much heat as it receives from the Sun, so these layers may be warm enough to sustain floating microorganisms if chemical evolution has progressed that far. The Galileo probe, which is scheduled to be released from the Jupiter orbiter in March 1985, will report on the composition of Jupiter's atmosphere, its temperature and pressure, and even detect lightning flashes as it falls, but it will not carry any life sensors. Saturn, Uranus, and Neptune may be as favorable or as hostile to life as Jupiter. Pluto, however, is assuredly a frozen waste that would make Antarctica in midwinter seem mild by comparison.

With the progress of space exploration, the weight of evidence is piling up, and the balance now tilts towards the Earth's being the only site of life in our solar system. If, after all the planets, their moons, and the larger asteroids have been investigated by unmanned landers and by atmospheric probes, and the Earth remains as the only preserve of life, should we be discouraged? Not at all. Absence of life in the solar system means that we need not worry about alien forms of diseases attacking the explorers or their plants and animals. The solar system will, therefore, be wide open to colonization.

Life Outside the Solar System

What are the chances that life exists outside the solar system? This is a subject for speculation, for we have little evidence to go on. Many scientists believe that life is not uncommon in the universe, while intelligent life is much rarer than lower forms of life. Dr. Frank D. Drake of Cornell University devised an equation for estimating the possible number of civilizations (not just life-suitable planets) in the Galaxy:[2]

$$N = R_* f_p n_e f_l f_i f_c L \tag{1}$$

where

N = number of civilizations in the Galaxy

R_* = rate of star formation (number of new stars formed each year) in the Galaxy

f_p = fraction of stars in the Galaxy that have planets

n_e = average number of planets in these planetary systems that are suitable for life

f_l = fraction of life-suitable planets on which life starts

f_i = fraction of these planets on which intelligent life develops

f_c = fraction of these planets on which intelligent forms reach the stage of interstellar communication

L = average lifetime of these civilizations

Of all these factors, only one, R_*, has any basis in observation, since our knowledge of even the nearest stars is pitifully scant. In our Galaxy of some 100 billion stars the current rate of star formation is estimated at ten per year. It was probably higher in the past. The rest of these factors are, at best, only educated guesses. There is no firm evidence for the existence of even a single planet outside our own solar system.

Drake's equation suggests the direction in which science must move to learn what lies beyond our little solar system. It is a guide both to the search for extraterrestrial intelligence (SETI) and to the detection of extrasolar planets. It begins with astronomical data and works onward to Galactic history.

Let us split Drake's equation as follows:

$$N = FT \qquad (2)$$

F = formation rate of life-suitable planets
T = fraction of life-suitable planets that now have advanced civilizations, or the average lifetime of a civilization divided by the number of suitable planets

Then:

$$F = R_* f_p n_e \qquad (3)$$
$$T = f_l f_i f_c L \qquad (4)$$

Let us consider the formation of life-suitable planets [equation (3)]. We expect that the number of planets suitable for life will be far greater than the number of advanced civilizations. If there are 100 billion stars in our Galaxy, and one-tenth of these stars have planetary systems, and one-tenth of these planetary systems have on the average one planet suitable for life, then there are a billion planets in the Galaxy fit for life

The division of Drake's equation into two components reveals two opposite approaches to investigating the existence of life in the universe. We can start at the base, improve astronomical techniques, and search for planetary systems. This approach will yield dependable results, whether positive or negative, and will accumulate knowledge of extrasolar systems as astronomy advances. It will help us prepare for the day when the first starship leaves on the grandest voyage of history, its crew equipped with knowledge of what to expect at its destination. Or we can begin at the clouded summit, build bigger and better radio-telescopes, and improve our techniques until, hopefully, one day an unmistakable signal is received. While antithetic, these approaches are neither exclusive or contradictory. They supplement each other, and both should be pursued much more vigorously than they are today.

In 1908 the distinguished Swedish chemist, Svante Arrhenius, theorized that life on Earth had been started by spores from a distant stellar system swept outward by the pressure of starlight. Most scientists have dismissed panspermia, as Arrhenius called it, since it shifts the problem of the origin of life from Earth to another solar system. Dr. Francis Crick[3] of the British Medical Research Council and Dr. Leo

M. Orgel of the Salk Institute have stated that such a journey is impossible, since cosmic rays would kill the spores in the course of the centuries it would take them to reach the Earth. But this offhand rejection may not be warranted. A modification of the theory, called directed panspermia, is not ruled out.[4] Let us look more closely at this modified theory, as advanced by Crick and Orgel.

Life may be eons old. The Sun is five billion years old, while our Galaxy is about 13 billion years old. The first generation of stars formed in the Galaxy was composed almost entirely of hydrogen and helium, and so it lacked an entourage of solid planets. The more massive first-generation stars produced the heavier elements in their depths through nucleosynthesis; eventually these stars exploded, enriching interstellar space with the heavier elements. The second-generation stars therefore had a higher percentage of heavier elements. In some favorable regions of the Galaxy the heavy-element density may have increased sufficiently that solid planets with Earth-like dimensions may have been created two billion years after the Galaxy's birth. Therefore, Crick and Orgel argue, Earth-like planets may have existed six billion years before our own solar system was formed.

Life on Earth began about 3.4 billion years ago. Since extraterrestrial civilizations may have existed at that time, one of them may have decided to dispatch starships to fertilize life-capable planets throughout the Galaxy. The starships would have provided protection against radiation, and they could easily have kept the microorganism cultures viable for long periods, possibly for millions of years, at temperatures close to absolute zero in interstellar space.

This is more like science fiction than science! Can it be proven? One way would be to demonstrate that life on Earth has properties that are inexplicable if life began on Earth. These properties must characterize all species from microorganisms to humans.

Dr. Paul Becquerel in 1950 did an experiment to study the possibility of life's surviving in space.[5] He immersed spores of several species of bacteria and fungi for two hours in a very high vacuum at a temperature just above absolute zero. When returned to normal conditions, the spores germinated. Apart from the danger of cosmic radiation, then, it seems that spores could survive in space. This investigation should be followed up by exposing spores for a long period in space, where they would be subjected to cosmic rays.

The case for directed panspermia gains plausibility if plants, animals, and bacteria contain elements that are extremely scarce on Earth but relatively abundant elsewhere in the universe. Molybdenum may be such an element. Some stars are rich in molybdenum relative to the Sun, while molybdenum makes up only 0.2 percent of the Earth's mass.

Crick and Orgel were very cautious in advancing the theory of directed panspermia. They proposed extraordinarily diverse studies to test it, such as analyzing the design of long-range starships and their missions, the lifespan of organisms at near absolute zero, and the growth of living microorganisms such as bacteria and viruses in media imitating the presumed primeval Earth atmosphere. One point stands out: if it could be shown that Crick and Orgel are right, the existence of life in the universe beyond Earth would be proven.

Criticism of the Crick-Orgel hypothesis was not long in coming. Dr. Thomas H. Jukes of the Space Science Laboratory, University of California at Berkeley, objected that besides molybdenum, enzymes contain five other trace elements—iron, zinc, copper, manganese and selenium, which have a concentration in sea water of about one part per billion.[6] What, then, was so unusual about their containing molybdenum?

W. R. Chappell, R. R. Meglen, and D. D. Runnels, of the Molybdenum Project, University of Colorado, pointed out that the average abundance of an element on Earth does not mean much in itself.[7] What is more important is its availability, especially its abundance in sea water, since life presumably began in the ocean. Chromium and nickel, which are reasonably abundant on Earth, do play a role in biochemistry, Crick and Orgel to the contrary, and hence are important to life along with molybdenum. Moreover, it is difficult to distinguish the natural background level of trace elements from pollution.

Orgel replied that if it could be demonstrated that the ocean contained molybdenum when the young Earth had a reducing atmosphere, then he and Crick were wrong, but that had not been shown.[8] Furthermore, neither chromium nor nickel plays a role in the biochemistry of microorganisms or in the major biochemical functions of higher organisms. Molybdenum, in contrast, is a major metal cofactor for several important enzymes.

Later, Dr. Devlin M. Gualtieri of the University of Pittsburgh proposed a general test of directed panspermia that reformulates an idea of Crick and Orgel.[9] "If," he wrote, "the distribution of elements in terrestrial organisms differs greatly from that of their environment, or if several elements are unaccountably enriched or deficient, these hypotheses may be justified." Many life forms concentrate elements. For example, some seaweeds are far richer in iodine than the sea water in which they live. Multicellular plants need boron, yet fungi and green algae do not. However, the concentration of trace elements in bacteria, fungi, plants, and land animals agrees well with their concentration in the ocean, their "genesis environment." Gualtieri maintained that the evidence supports the belief that Earth life began in the oceans of the primeval Earth.

While it appears improbable, the directed-panspermia theory is by no means impossible. It may well remain suspended in a state of nagging abeyance until someone comes up with a decisive test.

There is a faint possibility that evidence of a relic of extraterrestrial intelligence in our own solar system could be detected with the giant 300-meter (1000-foot) Arecibo radiotelescope—a huge metal bowl set among the hills of Puerto Rico that can view the planets and their moons. Also, it has been equipped to operate as an extremely powerful radar.

Dr. Claude Anderson of the University of California at Berkeley has suggested a radar search of the outer bodies of the solar system for bright pointlike radar echoes.[10] He believes that the solar system may have been visited at least once in its four billion years and that the visitors may have left a memento of their stay among the outer planets, where they extracted deuterium for their starship engines. The relic could be a corner reflector, possibly a kilometer across. This type of reflector would reflect radar beams right back to the transmitter regardless of the direction in which the reflector was pointing. At the distance of Neptune, a reflector less than a kilometer across would have the same reflectivity as the planet Mercury, which is 4880 kilometers (3030 miles) in diameter. This proposal is worth trying, for the results would definitely be positive or negative. Detecting a bright point on or near Ganymede, say, would be exciting.

The Origin of Life

If life began independently on Earth, how can we discover whether there is life elsewhere in the universe? Theory, unsupported by research, is useless. Fortunately, we may pursue several lines of attack in the laboratory and the field. Exobiology, as the new interdisciplinary field is called, investigates the origin of life and its existence outside the Earth. Exobiological investigators include paleontologists, geochemists, biochemists, geophysicists, geologists, and astrophysicists.

Life can exist only in a suitable environment—a truism that is ignored by some of the more extreme speculators on life in the universe. However, even the most life-favorable planet may not necessarily shelter algae or even viruses. Terrestrial planets with oxygen-rich atmospheres, the oxygen derived from dissociated water vapor from a central sun, may boast warm oceans that are inviting, lovely, and sterile.

The Earth is about 4.7 billion years old. The oldest known fossils are the microscopic blue-green algae of the Fig Tree shale beds of South Africa. They are about 3.5 billion years old. Dr. Cyril Ponnamperuma of the University of Maryland believes that the newly formed, fiery Earth took about 500 million years to cool enough so the vast clouds of steam erupted by volcanoes could condense and run down into the great basins to form the first oceans.[11] Then, 4.2 billion years ago, the stage was set for life to begin. The Fig Tree algae may have been the climax of 500 million years of evolution from lifeless matter. Ponnamperuma believes that life could have developed from lifeless compounds very quickly: "Indeed, if the right components and conditions existed in the primordial ocean, molecular organization into a living entity would have been instantaneous!"

Whether it was instantaneous or took 100 million years, the crucial step from the complex organic molecule such as a protein to the first living organism is that the organism, no matter how simple, can reproduce itself. That is, it can make copies of itself. Edward Argyle of the Dominion Radio Astrophysical Observatory believes that this transformation from lifeless matter to life could not have been due to chance, since the gap is too great.

After the Earth's crust solidified and the oceans were formed, sunlight, lightning, and other forms of free energy are believed to have induced the synthesis of life-related compounds such as amino acids, purines, sugars, and pyrmidines in the water, which may have become a thin organic soup. But none of these compounds, no matter how complex, could have developed into the first living cell, for they lacked the information content needed for reproduction. The viruses, the simplest living things, are only half alive, since they can not live outside live cells. They enter the cell, take over its genetic machinery, and direct it to make copies of themselves. Viruses have about 50 genes that contain a total of 120,000 bits of information. (The binary digit or bit is the information unit. It stands for two alternate states such as "on-off" or "yes-no.") The common bacterium *Escherichia coli* has 2500 genes that contain altogether 60 million bits of information that direct its reproduction and other activities. *E. coli* reproduces very readily.

Argyle believes that if the first living cell were as complex as *E. coli*, purely random chemical reactions between the various compounds in the water would have had to create a million bits of information and keep them together, all by chance. The origin of life by this route is almost impossible, even if there were a billion Earthlike planets in the Galaxy with prebiotic environments identical to the Earth's. Over a period of 500 million years the average amount of information that could have been incorporated into organic molecules in this way would have been only about 200 bits.

The evolution of life has been speeded up by the mutation of genes, the natural selection of favorable mutations, and their retention by subsequent generations. In this way, humans have come into being with 100,000 genes, or 240 million bits of information in each of their cells.

Argyle cites the analogy of a door to illustrate how Darwinian evolution, the evolution of life, works. Imagine a door that can be opened only by correctly setting a sequence of switches. The door opens a little when a switch is set correctly. The operator can quickly open the door by watching what happens when he sets a switch. All he has to do is to set a switch and change its setting. If the door opens a little, he leaves that switch alone, tries another, and so continues until the door is wide open. If the switch does not open the door a little, he returns it to its former state. That is, if the switch was "on" and he has turned it "off", he would turn it back to "on."

The door stands for the reproducing species, and the width of its opening stands for its biological success. Each trial of a switch represents a generation. The two states of the switch stand for the two cells formed by the division of a parent cell, one normal and one mutant. The operator is the environment, which destroys the "unfit" states and leaves the "fit" states or cells to serve as the parents of the next generation.

Consider what this analogy implies. Suppose that the door were kept closed by twenty independent switches, all having to be set correctly to allow the door to be opened at all, like a combination lock. One million trials would have to be made to get the door to open. By Darwinian evolution, as we have seen, only 60 trials or generations, not a million, would be needed to produce the same change in a species. Darwinian evolution enormously speeds up the rate of accumulating information bits and so the progress of life. Argyle said:

> *The Darwinian organism acts like a machine for generating information. Its special function is to copy all its genes, including those carrying random alterations to the message. All new messages are then subjected to environmental scrutiny. It is immaterial that only rarely does a new message pass the test. Once approved, it is copied at an exponentially increasing rate and for a time, becomes the "standard" message that underlies future attempts to encode even more information about the environment's tolerance for life. This process works because the reproductive power of a population of organisms exceeds the environmental culling that takes place between generations. The cost imposed by genetic experimentation is paid out of surplus reproduction. Meanwhile mutations that are not rejected add information to the genes at the Darwinian rate.*[12]

How was the giant advance made from complex organic compounds to the simplest possible organism? Suppose, Argyle said, that these compounds floating in the water could make copies of each other but not of themselves. They would then make up a reproducing chemical community, probably in shallow water near the shore. The compounds might be held in tiny interstices in the clay on the sea floor. In time, groups of reproducing compounds would be formed. They would accurately reproduce themselves, since the genes in these groups would have accumulated a million or more bits of information. Life would have begun.

If this process occurred on Earth, it could occur anywhere in the universe on other planets under similar conditions. Life may not be common in the universe, but it may be widely dispersed.

Radio astronomy provides another line of research on the problem of life in the universe. Interstellar molecules selectively absorb radio waves emitted by stars and galaxies, so the radioastronomers can identify the substances thinly scattered through the depths of space. More than fifty have been identified so far—not only the simpler ones, such as ammonia and water, but also more complex molecules that are important in chemical evolution, such as formaldehyde, methanol (wood alcohol), and methylamine. There is a possibility that even larger molecules such as the porphyrins, building blocks of life, may yet be found in interstellar space.

Comets, which are believed to be relics of the orginal solar nebula, are rich in cyanides that are among the key chemicals of chemical evolution. Also, one kind of meteorite, the type 1 carbonaceous chondrite, has a considerable amount of carbon, mostly as organic compounds. Dr. Edward Anders of the University of Chicago, who has done extensive research on the chemistry of the solar system, says that they are the most primitive known sample of matter.[13] Chemists have determined that the carbonaceous chondrites solidified from the nebular gas at about 87°C (170°F) at a pressure of about 4 one-millionths of an atmosphere, possibly within a few thousand years.

The Earth and the other inner planets may have received much of their organic matter early in the formation of the solar system when they were bombarded by small carbonaceous meteorites ranging from about a centimeter to a meter in diameter. In this way the Earth may have gained millions of tons of "ready-made" organic compounds, which would have speeded up the process of chemical evolution leading to life. The Murchison meteorite, for example, contains definite traces of amino acids—glycine, alanine, valine, proline, aspartic acid, and glutamic acid, constituents of proteins. The transition from prebiotic matter to life may have been greatly accelerated in this way.

Returning to Drake's equation, let us look at its second half, which is concerned with the search for extraterrestrial intelligence:

$$T = f_l f_i f_c L \qquad (4)$$

where

T = fraction of life-suitable planets having at present advanced civilizations
f_l = fraction of life-suitable planets on which life starts
f_i = fraction of these planets on which intelligent life develops
f_c = fraction of these planets on which intelligent life reaches the stage of interstellar communication
L = average life of these civilizations

The Search for Extraterrestrial Intelligence

Dr. Carl Sagan of Cornell University believes that about a million civilizations exist in the Galaxy on the same or a more advanced level than Earth civilization. He estimates that on the average they are about 300 light-years apart. How fascinating it would be if we could be certain that there was another civilization 300 light-years away! Drake's equation cannot help us here: its latter part is comprised of completely unknown factors. In short, the Search for Extraterrestrial Intelligence (SETI) is an act of faith.

SETI began in September 1959 when Drs. Philip Morrison and Guiseppe Cocconi published a short paper in *Nature* in which they proposed attempting to detect radio signals from other civilizations. The first attempt was made by Drake with the 85-foot radio telescope of the National Radio Astronomy Observatory at Green Bank, West Virginia, situated in a valley well protected from radio interference by the Blue Ridge range. The telescope was directed at two stars, Tau Ceti and Epsilon Eridani. The receiver was tuned to the frequency of 1420 megahertz with a bandwidth of 100 hertz. (A hertz is a cycle per second and a megahertz is a million cycles per second.) The search began on April 8, 1960, and continued for three months with a total observing time of 150 hours. No signals were detected. Since then, more than a score of searches have been made, both in the United States and the Soviet Union, with a total of more than 2000 observing hours—again without success.

If extraterrestrial civilizations exist, why have they not been detected after decades of searching? Is SETI the pursuit of a phantom? Or are existing extraterrestrial civilizations all radio silent? Or is the nearest extraterrestrial civilization so distant that its radio signals cannot be retrieved from the celestial static? To begin to answer these questions, let us look at the constraints that the radio astronomers must somehow master if SETI is ever to succeed.

The difficulties are gigantic, much greater than many of the SETI enthusiasts will admit. To start, the signal may come from any direction, so the whole celestial sphere should be scanned to detect it. However, a radio beam from a distant solar system will cover a very tiny area of the order of a square arc-second. For comparison, the total area of the celestial sphere is 535 billion square arc-seconds! The searching radio telescope should be able to point in 535 billion directions in succession. Any signal that would be detected, moreover, would far more likely be accidental than deliberately aimed at the Earth and so probably would be intermittent. Furthermore, to be absolutely certain, the entire electromagnetic spectrum from gamma rays to very long radio waves miles long should be continuously monitored, a frequency range of 1×10^{21} cycles. Impossible? No, but extraordinarily difficult.

How could an artificial radio signal be distinguished from the radio static in which it would be immersed? Drake set forth several criteria that have been generally accepted.[14] First, the signals can be expected to have a narrow bandwidth or range in frequency. This is reasonable in terms of communications theory, which states that, for a signal at a set power level, the narrower the bandwidth, the greater the range. Most natural sources of radio noise such as the Sun, stars, and galaxies emit very broad outbursts with bandwidths of thousands of hertz, so this criterion should be helpful. Some natural sources emit narrowband radiation, in the form of spectral lines in the radio region. These wavelengths are well known, are predictable in principle, and thus can be avoided for SETI searches.

Second, the artificial signal would likely be Doppler-shifted, as it would be sent from a transmitter in a stellar system. The Doppler shift is the change in frequency of a signal as the source approaches or recedes from the receiver. A common example of the Doppler shift is the familiar drop in pitch of an auto horn as a vehicle passes by. If the artificial signal were sent by a spaceship in orbit around a star, it would have a single Doppler shift pattern. If it were being sent from a transmitter on a planet orbiting the star, it would show a Doppler shift due to two motions, one caused by the revolution of the planet around the star and the other by its rotation on its axis, provided the geometry was favorable.

Third, the signal should be modulated; that is, it should vary with time in some sort of pattern. A signal that does not change with time cannot convey any information. Signals used for communication such as radio and television are modulated. Finally, the signal should come from the direction of a star.

Let us turn the problem around. How could an extraterrestrial civilization detect the Earth at radio wavelengths? By pointing their supersensitive radio telescopes at the Sun, their radio astronomers would detect the Sun's static and the deep radio growls of Jupiter. Amid the noise, they would find narrowband signals that would appear and disappear with marked regularity every 24 hours: your favorite daily TV program and the BMEWS (Ballistic Missile Early Warning System) radar signals, for example. By following this fascinating phenomenon for a long time, they would find that the frequency was very smoothly and regularly changing over a period of 365¼ days. Moreover, the bandwidths of the signals would be neatly arranged through the radio spectrum. With this evidence before them, even their most skeptical radio astronomers would be convinced that the signals were coming from a culture residing on a planet that was circling a star in a period of 365¼ days and turning on its axis once every 24 hours.

Ever since radio broadcasting began in the early 1920s, our civilization has unintentionally been spreading radio and more recently television signals through interstellar space. By now, our "leakage" can be detected at least 30 light-years away, a short distance in cosmic terms but further than many of the nearest stars such as Alpha Centauri, 4.3 light-years away; Sirius, 8.6 light-years; Tau Ceti, 10.2 light-years; and Epsilon Eridani, 10.5 light-years.

In the summer of 1971 the NASA Ames Research Center, in cooperation with Stanford University and the American Society for Engineering Education, conducted a summer-long study project, Project Cyclops, whose purpose was: "To assess what would be required in hardware, manpower, time and funding to mount a realistic effort, using present (or near-term future) state-of-the-art technique, aimed at detecting the existence of extraterrestrial (extrasolar system) life."

The problem Project Cyclops examined—that of locating an intelligent signal amid the radio noise—is like looking for a needle in a haystack without being sure that the needle is there. The presumed signal will certainly be very weak. Also, it may be detected for only a short time as the Earth rotates, and the transmitter may also be revolving to make it even more complicated. Perhaps the signal would be pointing at the Earth for only a second each day.

Most radio astronomers agree that the best region of the electromagnetic spectrum for detecting signals is the microwave band. Microwaves are the short radio waves. The quietest region in the radio spectrum is the free-space microwave window, between about 1 to 60 GHz (one billion to 60 billion cycles per second). Here on Earth, the water vapor and nitrogen in the atmosphere absorb radio waves between 22 and 60 GHz, in effect blocking the window above 10 GHz. The radioastronomers therefore believe that 1 to 10 GHz would be a good region for the search, or at least to get started.

Restricting the search to 1 to 10 GHz reduces the size of the problem but not by much, for ten billion 1-Hz channels would have to be searched. If, with Dr. Bernard M. Oliver of Hewlett-Packard, one of the two codirectors of the study project, we suppose that the signal may happen to be aimed at the Earth for one second each day, then we would have to search each channel for about 31 hours (100,-000 seconds) to accumulate enough of the weak signal to reliably detect it. Suppose that the entire celestial sphere was searched with a receiver having only 1 Hz bandwidth, then the search would take[15]

$$\begin{aligned}
Time &= 10^{10} \text{ (directions)} \times 10^{10} \text{ (channels per direction)} \times 10^5 \text{ (seconds per channel)} \\
&= 10^{25} \text{ seconds or } 3 \times 10^{17} \text{ years!}
\end{aligned}$$

This is 30 million times the age of the Galaxy!

Oliver proposed a strategy to reduce the search period to a rational length of time: namely, do not search all of the sky, but rather concentrate on those stars somewhat similar to the Sun that are within 1000 light-years. There may be about a million such stars. If possible, while observing a star, scan all of the frequency channels at the same time. Finally, set up an arbitrary but reasonable criterion: the signal we are looking at either comes from a nearby star or is on all of the time. This reduces the search time to about 50 years. If the cooler K-type stars that are unlikely to have planets are omitted, the search time becomes about 25 years.

Unfortunately, astronomers have identified only a small fraction of the stars within 1000 light-years of the Sun, as the process of determining the distance of stars is laborious and time consuming. Those nearby stars have been discovered are the ones that are bright, peculiar, or have large motions across the celestial sphere. Advances in technology will help in the discovery of additional nearby stars.

The outcome of the summer study was Project Cyclops, a grandly imaginative scheme that would be a great advance in radio astronomy even if it did not find an artificial signal for a century after it went into operation. It would not be a single gigantic radio telescope. Detecting extraterrestrial signals most likely requires an antenna with a diameter of about 5 kilometers (3 miles), and such a steerable radio telescope is out of the question on Earth. A giant radio telescope in space may be feasible but only with advanced technology. A practical approach would be to construct arrays of standard, reasonably sized antennas that could work together to form a giant antenna.

Project Cyclops would, therefore, be made up of 100-meter (330-foot) radio telescopes covering many hectares (acres). In the center of the field would be a large control building. A network of underground utility tunnels would radiate to the telescopes. Beginning with a score or so of antennas, Project Cyclops would grow to a forest of possibly 2500 antennas, 16 kilometers (10 miles) across.

The cost and the range would depend upon the size. At 10 kilometers diameter (6 miles), Project Cyclops would number 1000 antennas equal to a single antenna 3160 meters (13,000 feet) in diameter and be four million times as sensitive as the radio telescope used by Drake for Project Ozma, the pioneering SETI attempt. It could detect a signal coming from 20,000 light-years and search the radio waves 200,000 times as fast as Ozma.

Comparison of Project Cyclops with Project Ozma[16] is instructive. If a radio transmitter at the distance of Tau Ceti, 10.2 light-years away, had been detected by the Project Ozma recorders, it would have had to transmit with a power of ten billion kilowatts. American radio stations broadcast mostly with a power of around 50 kilowatts. Project Cyclops could detect a Tau Ceti transmitter if it broadcast with a power of 2500 kilowatts.

Even if Cyclops never discovers an intelligent signal, it will be rewarding. We are venturing very timidly into the universe, but we

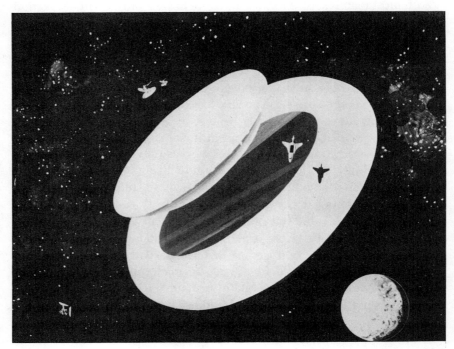

Figure 2-1. A radio antenna in space. An artist's concept of a 300-meter (1000-foot) SETI (Search for Extraterrestrial Intelligence) antenna in space. The picture shows two feeds, a relay satellite, a radio frequency interference shield behind the antenna, and a space shuttle. Such a radio telescope would be located in geosynchronous orbit (35,900 kilometers or 22,300 miles) about the Earth or higher up. (Courtesy of NASA)

cannot turn back. The absence of extraterrestrial civilizations in nearby solar systems will be very important to space explorers and very significant. Cyclops, too, will be a marvelous astronomical instrument that will enormously advance space science. Used as a radar, a 3-kilometer-equivalent Cyclops would have a combined area of seven million square meters (70 million square feet) from 900 antennas, each 100 meters across. It would be far more powerful than the Jet Propulsion Laboratory's 65-meter (210-foot) radio telescopes that have been used to map Mercury and Venus by radar. It could map the whole solar system out to Pluto. Working with radio telescopes 4800 kilometers (3000 miles) away, it could measure the direction of stars to a hundredth of an arc-second. In fact, it could surpass our largest optical

Figure 2-2. A SETI system on the Moon. An artist's concept of an array of three large radio antennas constructed within craters on the far side of the Moon and connected to form a SETI system. (Courtesy of NASA)

telescopes in precisely determining the directions of stars. With a baseline of, say, 16,000 kilometers (10,000 miles), it could measure the diameters of stars to within a ten-thousandth of an arc-second. (Large optical telescopes now being considered could do even better, especially if located in space.) With such precision, the trajectories of nearby stars could reveal the pull of their Jovian-sized planets, if they had any. Such stars, needless to say, would get special attention from the Cyclops staff.

Since 1972, when Project Cyclops was proposed, advances in the technology applicable to radio astronomy show that great improvements in sensitivity will make it possible to do meaningful SETI programs with smaller antenna areas. This would greatly decrease the cost of Cyclops.

Notes

1. Galileo Galilei, *Dialogue Concerning the Two Chief World Systems*, 2nd rev. ed., trans. Stillman Drake (Berkeley: University of California Press, 1967), p. 58.
2. *Possibility of Intelligent Life Elsewhere in the Universe*, rev. ed., House Committee on Science and Technology, 95th Congress, 1st Session, (Washington, D.C.: Government Printing Office, October 1977), p. 29.
3. Dr. Francis H. C. Crick is the famous biologist who, with James Watson, discovered the spiral structure of DNA, the bearer of heredity in the living cell.
4. F. H. C. Crick and L. E. Orgel, "Directed Panspermia," *Icarus*, 19 (1973), 341–346.
5. P. Becquerel, "La Suspension de la Vie des Spores des Bactéries et des Moisisures desséchées dans le Vide, vers le zéro absolu, Ses conséquences pour la dissemination et la conservation de la Vie dans l'Univers," *Comptes Rendus*, 231 (1950), 1274.
6. T. H. Jukes, "Sea-water and the Origins of Life," *Icarus*, 21 (1974), 516–517.
7. W. R. Chappell, R. R. Meglen, and D. D. Runnels, "Comments on 'Directed Panspermia'," *Icarus*, 21 (1974), 513–514.
8. L. E. Orgel, "Reply: Comments on 'Directed Panspermia' and 'Seawater and the Origin of Life'," *Icarus*, 21 (1974), 518.

9. D. M. Gualtieri, "Trace Elements and the Panspermia Hypothesis," *Icarus*, 30 (1977), 234–238.

10. C. W. Anderson, "A Relic Interstellar Corner Reflector in the Solar System," *Mercury*, 3 (1974), 2–3.

11. C. Ponnamperuma, "Life Beyond the Earth," *Astronautics and Aeronautics*, 14 (November 1976), 50–54.

12. E. Argyle, "Chance and the Origin of Life," *Origins of Life*, 8 (1977), 287–298. Reprinted by permission of D. Reidel Publishing Company.

13. E. Anders, R. Hayatsu, and M. H. Studier, "Organic Compounds in Meteorites," *Science*, 182 (1973), 781–789.

14. H. G. W. Cameron, ed., *Interstellar Communications* (New York: W. A. Benjamin, 1963), p. 170.

15. *Project Cyclops* (CR 11445, NASA Ames Research Center, 1972), p. 58.

16. Cameron, *Interstellar Communications*, p. 176.

3
SURVEY OF THE SOLAR SYSTEM

One of Voltaire's stories is about the adventures of Micromegas, a giant who visited the solar system in the 1700s. If the grandson of Micromegas were sent in the 1980s to update his grandfather's report, he would be told very sternly to make a fast survey and not to waste too much time on it, since, according to his grandfather, the system was a mediocre affair. A popular version of the grandson's technical report follows:

The Sun is a self-luminous sphere of gas that has a mass of 1.99×10^{33} grams (2.1×10^{27} tons), a radius of 696,000 kilometers (432,600 miles), and a density of 1.4 gm/cm³ that is 1.4 times that of water. The apparent surface or photosphere is the layer about 200 kilometers (125 miles) thick where the gases become opaque. The surface temperature is about 5600°K (5300°C, 9600°F). In the photosphere, the temperature and the density decrease as one moves outward. The escape velocity from the solar surface is 617 kilometers (384 miles) per second. The chemical composition of the surface layers is, by number of atoms, about 90 percent hydrogen, 9 percent helium, and 1 percent heavier elements.

Outside the photosphere lies the chromosphere, which is about 8000 kilometers (5000 miles) deep. Here the temperature increases and the density decreases as one moves upward. The chromosphere merges into the outermost layer, the corona, which is so hot that it is unstable. The coronal gases expand into the inner solar system to form the solar wind with speeds of at least 500 to 1000 kilometers (300 to 600 miles) a second. The solar wind reaches the realm of the Jovian planets, where it interacts with the material of interstellar space. The location of the boundary depends on the density of the interstellar matter and the intensity of solar activity. For average solar conditions and noncloud regions of the interstellar medium, it lies close to the orbit of Uranus.

The Sun, like trillions of other stars, is a fusion reactor. It is a main-sequence star that converts hydrogen into helium in its core. As one helium nucleus has slightly less mass than the four protons or hydrogen nuclei from which it was formed, the mass that is lost appears as energy. This is the energy that causes the Sun to shine.

The Sun has 99.86 percent of the mass of the solar system. The nine major planets have 0.135 percent, with Jupiter accounting for more than half of this share. The satellites or moons have the next largest share with 0.0004 percent. Comets total perhaps as much matter as the moons, while the asteroids or minor planets and the meteoroids each have about one-tenth as much. The interplanetary medium contains less than one-third as much as the meteoroids.

The planets form two sharply different groups. The four inner ones, Mercury, Venus, Earth, and Mars, are small, rocky spheres. Their atmospheres range from almost nonexistent for Mercury through partly transparent for Earth and Mars to opaque for Venus. The outer planets, Jupiter, Saturn, Uranus, and Neptune, are large with low density. They have massive atmospheres that probably extend downward to small rocky cores. The outermost planet, Pluto, is small and solid. While all the planets shine by reflected sunlight, Jupiter and Saturn and perhaps Uranus and Neptune radiate more energy than they receive from the Sun. This energy is due to very gradual contraction of their volume by gravity.

Table 3-1 presents data describing the physical properties of the nine planets. Note that their masses range from about one four-hundredth that of the Earth for Pluto to 318 Earth masses for Jupiter. All revolve around the Sun in the same direction, west to east, or counter-clockwise as observed from north of the mean plane of the solar system. Their orbital periods around the Sun range from 88 days

TABLE 3-1. Properties of the Planets

	Diameter			Mass Earth = 1	Semimajor Axis[a]			Orbital Period[b]
	Kilometers	Miles	Earth = 1		AU	Kilometers (millions)	Miles (millions)	
MERCURY	4880	3030	0.38	0.06	0.39	58	36	88 days
VENUS	12,110	7510	0.95	0.82	0.72	108	67	225 days
EARTH	12,760	7910	1.00	1.00	1.00	150	93	365 days
MARS	6800	4120	0.53	0.11	1.52	228	142	687 days
JUPITER	143,000	88,600	11.20	317.9	5.20	778	484	11.86 years
SATURN	121,000	75,000	9.49	95.2	9.54	1427	887	29.46 years
URANUS	47,000	29,200	3.69	14.6	19.18	2870	1785	84.01 years
NEPTUNE	45,000	27,900	3.50	17.2	30.06	4497	2797	164.8 years
PLUTO	2500	1550	0.19	0.002	39.52	5912	3670	248.4 years

Note: All figures are rounded off.
[a] The semimajor axis is half the long diameter of the elliptical orbit of the planet. It is equal to the planet's mean distance from the Sun.
[b] The planet's orbital period is the time it takes to make a complete revolution around the Sun. It is equivalent to its year.

for Mercury to 248 years for Pluto. Their average distance from the Sun varies from 58 million kilometers (36 million miles) for Mercury to 5.9 billion kilometers for Pluto (3.6 billion miles). To set the scale, the average distance from the Earth to the Sun is one astronomical unit. Mercury's average distance from the Sun is 0.327 AU, while Pluto's is 39.518 AU.

The Earth, the third planet from the Sun, is the only body in the solar system with extensive areas covered with liquid water. The Earth's surface is 29 percent dry land and 71 percent water-covered. The Earth lies in the middle of the zone favorable for sustaining carbon-based life. Of all the solid bodies in the solar system, it has the greatest surface gravity and, therefore, escape velocity. Hence, if intelligent life exists on its surface—and there are indications that it does, with a somewhat primitive technology—once these beings have acquired the technology to easily surmount the Earth's gravity, they will be able to visit the other interesting places on which they could land with relative ease, such as Mercury, Mars, the Moon and other satellites, the larger asteroids, and Pluto.

The Earth is a very pretty globe, showing large patches of blue and smaller ones of brown and green that appear through the breaks in the cloud cover, which at times conceals up to 70 percent of the planet's surface. The atmosphere at the surface consists of nitrogen (79%), oxygen (21%), and argon (1%) along with water vapor, carbon dioxide, and traces of other gases.

The average temperature of the Earth at sea level in middle latitudes is 286°K (13°C or 55°F) with a daily variation of about 11 K° (11 C° or 20 F°). The Earth's atmosphere and extensive oceans work together to keep its surface within a temperature range suitable for life. The atmosphere also filters the Sun's radiation and keeps out almost all the gamma rays, X rays, and ultraviolet light harmful to life. There are two major "windows" in the atmosphere—the optical window that admits the band of the electromagnetic spectrum to which our eyes are sensitive and the radio window in which the inhabitants broadcast a primitive kind of television.

The pressure and density of the Earth's atmosphere diminish with altitude. Half of the Earth's atmosphere lies below 5.5 kilometers (3.4 miles). The higher the altitude, the less air remains to serve as a shield against harmful radiation. The outer atmosphere, several Earth radii above sea level, gradually merges with the tenuous gas of interplanetary space, whose density is about 10 atoms per cubic centimeter. In contrast, the density of air at sea level is 2.7×10^{19} molecules per cubic centimeter.

Except for Pluto and its moon, Charon, the Earth has the largest satellite in the solar system relative to the size of its primary planet. The Moon's diameter is slightly over a quarter that of the Earth, its mass one eighty-first, and its surface gravity one-sixth. Its mean distance from the Earth is 384,404 kilometers (238,864 miles), and it revolves around the Earth once in 27.322 days. One hemisphere of the Moon always faces the Earth and the other, empty space, since it rotates on its axis with the same period in which it revolves around the Earth.

The most conspicuous surface features of the near side of the Moon are the maria—great plains with fairly smooth, flat floors that appear to be darker than the surrounding upland regions. There are only a few maria on the far side, which, like the entire surface, is studded with craters. There are fewer craters per unit surface area on the maria than over the rest of the surface, which suggests they are younger than the uplands. The craters range in size from Clavius and Grimaldi, which are about 240 kilometers (150 miles) in diameter, down to pits. Rays, which appear to be debris thrown out by the impacts that causes the craters, radiate outward from some of the larger, more distinct, and presumably more recent craters. There are also mountains, valleys, and rills. Since the Moon does not have an atmosphere, there has been little erosion, and the topography is rougher than on a body such as the Earth, which has an atmosphere.

Mercury, the closest planet to the Sun, looks much like the Moon with a heavily cratered surface, large basins resembling the lunar maria, rays, and many long cliffs. Owing to its higher surface gravity, the distribution of craters in the Mercurian uplands is somewhat different, as the material thrown out from the impacts hits the surface closer to the main crater than it does on the Moon. Mercury has a diameter of 4880 kilometers (3000 miles), which is 38 percent of the Earth's, and its noontime temperature goes up to 700°K (800°F) while the temperature at midnight may go down to 100°K (-285°F). Although a tenuous cloud of hydrogen surrounds the planet, Mercury does not have a permanent atmosphere. Its pressure is less than 10^{-11} that of the Earth's atmosphere.

Venus, the second planet from the Sun, is almost a twin of the Earth, having 95 percent of its diameter and 82 percent of its mass. But, unlike the Earth, Venus is blanketed by a thick cloud cover, and its surface atmospheric pressure is 89 times that of the Earth at sea level. The surface temperature is 725°K (850°F). The atmosphere is about 97 percent carbon dioxide, 3 percent nitrogen, and 0.13 percent water vapor, with traces of oxygen, argon, neon, and sulfur dioxide. The clouds contain sulfuric acid and other sulfur compounds. It is a very inhospitable place for life.

Mars, the fourth planet, orbits the Sun close to the outer boundary of the temperature zone for carbon-based life. It is a cold, dry desert planet with maximum daytime temperatures of up to 300°K (23°C, 74°F), while nighttime temperatures go down to 200°K (−73°C, −100°F). Mars is only a tenth as massive as the Earth, has a diameter about one-half of the Earth's, and has a surface gravity 38 percent of the Earth's. Even so, its surface area is comparable to that of all of the Earth's continents put together. Although the atmosphere has a surface pressure about 1 percent of the Earth's, it is dense enough to have a twilight zone. The atmosphere is mostly carbon dioxide with traces of other gases. Great dust storms at times sweep over the terrain, which has polar caps, wide plains, volcanoes, canyons, and craters.

Most of the asteroids or minor planets circle the Sun between Mars and Jupiter. A few, however, have orbits that cross the Earth's. Ceres, the largest asteroid, is 955 kilometers (600 miles) in diameter, but most asteroids are only a few kilometers in size. Perhaps 100,000 asteroids can be detected by a 1.2-meter (48-inch) Schmidt telescope like the one at Palomar Observatory. Some of them belong to families with similar orbits, suggesting they are results of collisions between larger bodies. There is certainly a continuous range in size from the smallest known asteroids down to the solid dust-sized particles, the micrometeoroids, that have been detected both by spacecraft and on Earth. The distinction between asteroids and meteoroids is simply whether the object can be detected by a telescope.

The four giant planets, Jupiter, Saturn, Uranus, and Neptune, are enormous compared to the inner planets, but are less dense and have values close to that of water. Since they have retained more primordial hydrogen and other light elements, their compositions are more like that of the Sun, which partly accounts for their low densities. They are the Jovian or gaseous planets, and all of them have very deep atmospheres.

Jupiter's diameter is about 11 times that of the Earth and it is 318 times as massive, making it the second largest body in the solar system. It is flattened by its rapid rotation, which averages 9 hours, 55 minutes. Jupiter is covered with colorful swirling clouds and bands. The darker bands are warmer than the adjacent light bands and apparently are at lower levels. The bands may be tinged red, yellow, orange, blue, blue-gray, brown, or light grey. The atmosphere extends down for perhaps 700 kilometers (430 miles), and below this level the hydrogen begins to liquefy. Further down, the hydrogen becomes a metallic liquid. Jupiter probably has an ice shell around a rocky core. Jupiter's magnetic field is stronger than that of any other planet, about ten times as strong as the Earth's. Voyager missions have also discovered that Jupiter has a ring like Saturn's, but is much fainter.

Saturn is a spectacular planet, as it is surrounded by a giant ring system. Apart from this, it resembles Jupiter, showing alternating light and dark cloud bands that are not as distinct as Jupiter's. Its equatorial diameter is 10 percent greater than its polar diameter, so it is markedly flattened. Although Saturn is 95 times as massive as the Earth, its average density is only 70 percent that of water. Its internal structure resembles Jupiter's. In the equatorial zone, it rotates once in only 10 hours 26 minutes, and tends to rotate more slowly with increasing distance from the equator. Saturn has a diameter of 121,000 kilometers (196,000 miles), and the rings extend from 13,000 kilometers (8000 miles) to 146,000 kilometers (100,000 miles). The rings are extremely thin and have gaps between them. They are made up of trillions of pebble-sized or smaller ice or ice-coated particles. Each pebble independently orbits the planet.

Uranus appears green, as it has a large amount of methane in its atmosphere. Its rings are far less extensive than Saturn's. The rotation axis lies almost in the orbital plane, which is unusual, as the other planets' rotation axes are within 25 degrees or so from the perpendicular to their orbital planes. Uranus is also about as oblate or flattened as Jupiter and has a slightly lesser density. It is only 14½ times as massive as the Earth and rotates once in 15 hours, 35 minutes.

Neptune resembles Uranus. It is similar in size but its density is greater, about 1.67 times that of water. Neptune's internal structure resembles that of Uranus and the other Jovian planets. Neptune rotates once in 19 hours, 35 minutes.

Pluto is the outermost planet. Its markedly noncircular orbit and small mass, about 0.02 percent of the Earth's, suggest that it was formed as a satellite of Neptune. The surface temperature is close to 40°K ($-387°$F), which is so low that almost all gases there are frozen. Thus it is unlikely that Pluto has an atmosphere.

All of the planets except Mercury and Venus have satellites or moons. The Earth has one; Mars, two; Jupiter, thirteen; Saturn, ten; Uranus, five; Neptune, two; and Pluto, one. Mercury and Venus lack moons, probably because they were formed close to the Sun. The

Jovian planets may have undiscovered smaller moons. From time to time, Jupiter may capture an asteroid and add it to its collection of moons or may lose to the asteroid belt a moon that is in a very large orbit.

The moons are a mixed group, ranging in size from Triton, 6000 kilometers in diameter (3700 miles), down to bodies a few kilometers across, such as the moons of Mars, Phobos and Deimos. Four moons, Triton, Titan, Ganymede, and Callisto, are larger than Mercury. These four plus Io, the Moon, and Europa are larger than Pluto. Titan has an atmosphere composed equally of hydrogen and methane, and its atmospheric pressure is at least four times that of Mars, although its surface temperature is close to 125K ($-150°C$ or $-240°F$).

Comets are the strangest members of the solar system. In one respect, they are among the largest bodies in the solar system, and in another, among the smallest. Their tails may stretch more than 150 million kilometers (100 million miles) and their heads may be larger than the Sun, yet they are mostly very tenuous gas. The solid nuclei in their heads, which generate the gas and dust, may be only 1 kilometer or less across. The nuclei are a mixture of small rocks and ices: water, methane, ammonia, and carbon dioxide. As the nucleus approaches the Sun and is warmed by its heat, the ices sublimate into gases. Comets are wasted away by their passages around the Sun, so they must eventually dissipate, except, perhaps, for a small solid nucleus. A large reservoir of comets beyond the orbit of Pluto replaces the comets that waste away.

Origin of the Solar System

The properties of all of these bodies—planets, moons, asteroids, and comets—have to be explained along with the structure of the solar system by a theory of its origin and evolution. The generally accepted theory is as follows:

About 4.5 billion years ago in an arm of our Galaxy the remnant of a supernova collided with the undisturbed region of the giant dust and gas cloud out of which its precursor star had been formed about a million years earlier. The shock waves penetrated some depth into the cloud and compressed it. Some portions of the cloud that had been denser than their surroundings now had enough mass and also were cool enough to collapse. One of these portions became the protosolar nebula. It had a mass somewhat greater than that of the solar system today.

Within a million years, the protosolar nebula collapsed to a radius of about 100 AU. It was relatively transparent until it reached this stage. Now, any photons (particles of light) emitted in the center were absorbed before they could escape into the surrounding space. The temperature of the solar nebula began to rise from a value of near 15°K as a result. Its density was only about 10^{-13} gm/cm^3. Since the inward pull of gravity tremendously exceeded the outward pressure, matter now fell freely toward the center.

In the next 100 years or so, the luminosity, surface temperature, and internal temperature continued to rise until the proto-Sun reached a visual luminosity about a thousand times that of the present Sun. The proto-Sun had a radius of 0.25 AU; then, the proto-Sun went through a brief period of adjustment, until the inward gravitational force was exactly equal to the outward pressure at every point within it. For the next million years the Sun slowly shrank while its surface temperature remained nearly constant. Its luminosity also slowly decreased, so that at the end of this period the Sun was slightly less luminous than it is today.

The direction of the Sun's evolution now changed. For the next 30 million years, while it continued to shrink, its surface temperature and luminosity slowly rose. Toward the end of this time it reached a sufficiently high central temperature so that hydrogen burning began in its core, slowing the shrinkage. Finally, the Sun was deriving all its energy from hydrogen fusion and none from shrinking. It had reached the zero-age main sequence. In the 4.5 billion years since then, the Sun has become 10 to 20 percent brighter, because of changes in its interior chemical composition as hydrogen fused into helium. This has had important consequences for the evolution of life.

In the next 5.5 billion years the Sun will brighten by a similar amount. This will slowly move the boundaries of the zone favorable to carbon-based life away from the sun. The Earth will still be suitable, while Mars will become more livable. Finally, when the Sun has used up the hydrogen in its core, its evolution into a red giant will have catastrophic consequences for life on Earth.

The solar system began as a rarefied, cool cloud of gas and dust whose chemical composition was the same as that of the Sun's surface layers today. Perhaps as much as 95 percent of the solar nebula formed the Sun. The remaining 5 percent collapsed, responding both to the proto-Sun and its own gravity.

The solar nebula, like almost every object in the universe, rotated about some internal axis. As it contracted, it spun faster. In time, collapse became impossible in its equatorial plane, where the nebula's rotation counteracted any tendency to collapse toward the proto-Sun. This occurred when the solar nebula was at least 40 AU in radius and perhaps slightly larger—about the stage when it became more opaque than transparent. However, nothing prevented collapse in the direction perpendicular to the equatorial plane, so it collapsed into a flat, circular disk whose thickness was determined mainly by gas pressure. The matter in the disk revolves in orbits described by Kepler's Laws (see Chapter 7).

The solar nebula was very turbulent; eddies of many sizes continually formed and dispersed. As the nebula contracted, its volume decreased; since its mass remained constant, its density increased. Finally, one of the eddies had sufficient mass for its size so that the mutual gravitational attraction of its particles was strong enough to prevent disruption by the tidal forces caused by the Sun. This large mass of dust and gas was the first protoplanet. As it moved around the proto-Sun in its orbit, it swept up material from the solar nebula. Soon other protoplanets began to form. Any that happened to be about the same distance from the Sun collided to form a single body, which tended to move in a circular orbit. Ultimately, there were eight protoplanets, each more massive than its descendants today.

Since the nebula was very dusty, the protoplanets were shielded from the proto-Sun. Except at the center, the solar nebula was cold. Ices condensed on the dust. Hence, each protoplanet consisted of a large mass of icy dust. As its mass increased, it slowly contracted. The heavier atoms moved toward the center, the lighter ones toward the exterior. At one stage every protoplanet, or at least the inner ones, probably rotated once on its axis as it went around the Sun. As they contracted, the protoplanets spun faster and became flattened disks of gas and dust. In the outer area of their disks, protosatellites formed, each moving in an approximately circular orbit. Mercury and Venus, however, did not spin fast enough to create satellites. The Moon may be a small protoplanet that was captured but not absorbed by the Earth, as its orbital plane is closer to the Earth's than to that of the Earth's equator.

The spacing of the planets from the Sun reflects in part the density of the solar nebula. The Sun either absorbed or expelled most of the material in the inner solar system, and the density of the solar system decreased toward the outer regions because of the way the system contracted. Consequently, the density was greatest in the central portion, where Jupiter and Saturn are located. The spacing also increased progressing outward, for otherwise the planets would have strongly interacted with each other. The largest protoplanets most likely had masses 1.5 percent of the Sun's.

The Sun slowly shrank as the solar system formed. Probably the solar wind was much stronger than it is today, and it gradually blew the dust away. When the space between the Sun and a protoplanet became sufficiently transparent, the protoplanet began to feel the effects of solar heating. Most of the mass of the protoplanets evaporated, especially hydrogen and helium. Thus, the inner planets came to consist of rocky material. The Earth, for example, lost all but 0.1 percent of the mass of the proto-Earth. Farther out, the more massive Jovian planets lost less, because of both their greater initial mass and their greater distance from the Sun. Jupiter retained perhaps 5 percent of its original mass.

The reason that Mars is less massive than the Earth is probably the disturbing gravitational effects of Jupiter. After proto-Jupiter formed, its powerful gravitational field prevented the formation of even a modest-sized planet in the asteroid belt. The protoplanets there could not combine, and the asteroids are their remnants. The comets were formed beyond the orbit of Neptune. When Pluto escaped from Neptune, it interacted with them and removed them to form a great spherical reservoir out to possibly a light-year or so. In the process many were lost, of course, to the solar system. Passing stars perturbed the orbits of some of the comets. A few had their orbits changed so that their perihelion points are close to the Sun.

Micromegas III concluded his report by noting disdainfully that this particular planetary system was typical and commonplace. He knew of much grander planetary systems, such as the one around Star Micromegas 3, which had been discovered by his grandfather. The rest of his report, unfortunately, has not been found. But see Table 3-1.

4
THE CONQUEST OF ZERO g

Before we can move into space en masse, two critical obstacles must be overcome: adequate space power and the handicaps of space existence. Space power is an engineering problem (discussed in Chapters 5 and 12). Space existence is a biological problem that must be solved for animals and plants as well as humans.

From the viewpoint of the aerospace engineer, space starts about 190 kilometers (120 miles) up, where a vehicle no longer can be supported by the air and no longer meets appreciable air resistance. If the vehicle reaches orbital velocity, it will circle the Earth and descend only when the feeble drag of the remaining atmosphere gradually slows it down. It imperceptibly falls to a lower orbit and eventually into the lower atmosphere, where it burns or breaks up in its plunge to Earth.

However, space does not suddenly start at 190 kilometers (120 miles), for there is a smooth transition from the lower atmosphere to the vacuum of space. Even at 3000 meters (10,000 feet) in mountain ranges the thin air affects breathing. One has to spend at least a night or more at this altitude to become acclimated. The mountaineers who climbed Mount Everest had to use oxygen masks. There are, then, areas on the Earth that have some of the features of space.

Human beings cannot survive in space without protection. The air pressure in our lungs is equal to the atmospheric pressure. As seen by medical science, space begins when one can no longer breathe in oxygen from the air. This is the 15,000-meter (50,000-foot) level, where the atmospheric pressure is only 87 millimeters of mercury compared to 760 millimeters at sea level. Here, a person without an oxygen mask on might just as well be in outer space as far as his or her chances for survival go.

At 19,200 meters (63,000 feet), air pressure is down to 47 millimeters, and even a person wearing an oxygen mask is in danger. Bubbles will form in the blood vessels, since the vapor pressure of water now equals atmospheric pressure. This is not the familiar boiling due to heating water. It is life-threatening, for the lungs will fill with water vapor. A pressurized cabin is necessary for survival.

This may not seem much of a problem, since millions of people have flown in airliners in pressurized passenger compartments. However, airliner cabins are not hermetically sealed. Powerful compressors pump in the thin outside air and keep the cabin at a tolerable air pressure, usually equivalent to 1500 meters (5,000 feet). As airliners fly higher for the sake of increasing speed, they will reach a height at which even the best compressors will not do the job, and sealed cabins will be mandatory. This occurs at 24,000 meters (80,000 feet). An airliner flying that high will have to carry a supply of oxygen and have a regenerating atmosphere control system like those on the Apollo and Skylab spacecraft. The transition from air travel to space travel may be less abrupt than the public believes.

Never before has life on Earth in all its four-billion-year history ventured into space. It is a new habitat for animals and plants as well as humans. Besides its almost total vacuum and weightlessness, space offers other formidable challenges: temperature that can abruptly switch from near absolute zero to over 100 degrees centigrade, cosmic rays, solar flares, and meteoroids.

Space temperatures have not been much trouble to aerospace engineers, since our technology can deal with them readily. The reflecting, absorbing, and reemitting of heat by the use of coatings, paints, and metal surfaces is well understood. Until now, engineers have been concerned mostly with keeping the temperatures of satellites and spacecraft down, not up—for people and equipment give off heat. Protection against heat loss is important for spacecraft and spaceships traveling to the outer solar system, but heat loss is easily blocked by insulation and the use of waste heat.

The Meteoroid Danger

In the early years of the Space Age, some astronomers feared that meteoroids would prove a menace to space flights, especially those that ventured away from the Earth to the Moon and planets. When space exploration became possible after World War II, scientists had to weigh the risks of encounter with meteoroids. Huge meteorites have struck the Earth and gouged out craters. The best-known crater, Arizona Meteor Crater, is 1250 meters (4100 feet) across and 180 meters (600 feet) deep. It was blown out by a meteorite 76 meters (250 feet) across that weighed a million tons.[1] The object struck the ground at 57,600 kilometers (36,000 miles) an hour with the explosive force of 30 million tons of TNT. The peak pressure of the impact was 15 million pounds per square inch!

The chance that a spaceship would be hit by such a monstrous object is very slight, but even a small meteoroid could do damage. Just what would happen if a meteoroid hit a spaceship? In 1946, Fletcher Watson of Harvard University estimated that about 4 percent of the rockets sent on flights to the Moon would be destroyed by meteoroids.[2] A significant fraction of the spaceships spending several hundred days en route to the planets would not succeed in evading the barrage of speeding meteoroids. A British scientist, N. H. Langton of the College of Rubber Technology, calculated that a meteoroid only 0.06 centimeter (0.025 inch) in diameter that struck the hull of a spaceship at the speed of 76 kilometers (47 miles) a second would heat the point of impact to 4 million degrees centigrade.[3]

Dr. Fred L. Whipple of the Smithsonian Astrophysical Observatory, a specialist on meteoroids and comets, was concerned enough to advocate that rockets should be equipped with meteoroid bumpers, thin sheets fastened around them to take the meteoroid impact and absorb the shock of the explosion.[4] The resulting vapor would spread harmlessly between the bumper and the rocket's skin. He estimated that a satellite 3 meters (10 feet) in diameter with a skin $1/3$ centimeter ($1/8$ inch) thick would be punctured about once every three weeks by a meteoroid about 0.2 centimeter ($1/12$ inch) across weighing about a ten-thousandth of a gram. This seemingly harmless particle would hit the satellite at a speed of 22 kilometers (14 miles) a second and pierce its skin.

Meteors have been observed from the ground for many years by photography and radar. NASA took the danger seriously and decided upon a survey. A series of six meteoroid satellites surveyed the frequency of meteoroids in space and determined the hazard they offered to space flight.

The space survey was undertaken because ground photography and radar could not give sufficiently accurate results. The size of the parent meteoroid of the photographed meteor is estimated from the brightness of its trail but with considerable uncertainty. Radar can detect meteors in daylight and through clouds, but it cannot directly determine the mass of the parent meteoroid. The best measurements have been those made directly by microphones (to detect acoustic impact) and thin metal sheet sensors (to detect perforations) that were mounted on meteoroid satellites.

The voluminous data received from the six meteoroid satellites showed that, while the possibility of a meteoroids hitting a spacecraft was far less than had been predicted, it was real nevertheless and should not be ignored. Whipple's bumper was adopted and named the meteoroid shield. NASA adopted criteria for meteoroids for use by engineers in designing protection against meteoroid impact.[5] The data can be summarized as follows: The range of meteoroids considered was from 10^{-12} gram to 1 gram. The density was taken to be 0.5 gram per cubic centimeter and the average velocity as 20 kilometers (12 miles) a second. The flux, or number of meteoroids passing through a square meter of space each second, decreases as the mass of the meteoroid increases. One meteoroid about a hundredth of a gram in mass passes through a square kilometer of space about once every 11 days, while a micrometeoroid a millionth of a gram passes through a square kilometer of space about once every 10 seconds. In contrast, a 1-gram meteoroid passes through a square kilometer of space about once every three years.

Space Radiation

Space radiation turned out to be a more serious danger. The first American satellite to go into orbit, which Russia's Premier Khruschev derisively compared to a grapefruit, discovered the Van Allen Belts. The radiation counters on Explorer I had been designed by Dr. James

Van Allen of the University of Iowa to observe possible radiation outside the Earth's atmosphere. They were brilliantly successful, for they detected radiation a thousand times as strong as the cosmic rays. The radiation in the Van Allen Belts is generated by particles from the Sun that hit the atoms in the upper atmosphere. The protons and electrons that are thrown off spiral around the lines of force of the Earth's magnetic field to form the Van Allen Belts.

There are two belts. The Inner Belt is not concentric with the Earth's surface, since it is 1480 kilometers (918 miles) high over Australia and 460 kilometers high (285 miles) over Brazil. The Inner Belt reaches to 5400 kilometers (3400 miles). The Outer Van Allen Belt extends from 16,000 kilometers (9900 miles) to 55,000 kilometers (34,000 miles) above the Earth. The Inner Belt is a danger zone. A single day in the Inner Belt would run the radiation dose for men in a spaceship above the tolerable daily limit. Space stations in low Earth orbit will therefore have to be located below 460 kilometers (285 miles) altitude. The Outer Van Allen Belt is much weaker than the Inner, for the energies of the protons in the Outer Belt are less than a million electron-volts. A geosynchronous space station circling the Earth at 35,000 kilometers (22,300 miles) would be safe. Also, a spaceship that passed through the Inner Belt in ten minutes would not be in any danger.

Beyond the Van Allen Belts, the space traveler faces the electromagnetic radiation of the Sun and cosmic rays as well as intermittent particle streams from solar flares. Since Explorer I, dozens of satellites have measured radiation near the Earth and through the solar system from orbits ranging from smaller than Mercury's to larger than Saturn's. The galactic cosmic rays from outside the solar system are not a serious hazard, even for voyages lasting more than a year. Neither is the solar wind, which flows in very thin streams from the Sun far out into the solar system. It is made up primarily of ionized hydrogen and helium and its speed reaches 1000 kilometers (620 miles) a second. While it is continual, the penetrating power of the solar wind is so feeble at one astronomical unit that it is not a hazard to an astronaut out in space in his space suit.

Solar flares are another matter. A solar flare is a sudden brightening of an area of the Sun in X rays, ultraviolet, and visible light. The flare ejects gas clouds that travel at 6000 kilometers (3700 miles) a second and reach the Earth in 24 hours. Major solar flares emit cosmic rays, streams of protons, and heavier nuclear particles with sufficient energy to penetrate metal.[6] A large solar flare can pour five million protons a second, traveling at nearly the speed of light, into a square centimeter (one-sixth of a square inch) of the Earth's atmosphere. Solar flares were first studied intensively during the International Geophysical Year of 1957. Thirty were detected during and shortly after the IGY. Six of them generated beams of particles with enough energy density to kill anyone in space in a space suit.

The radiation from a giant solar flare would endanger even astronauts in a spacecraft unless they were well shielded. Unprotected astronauts subjected to such flare radiation would feel tired and nauseous; they might vomit and lose their appetite. If the first giant solar flare were followed by a second one in a few days, their bodies would not have had time to recover from the first one.

In the late 1950s NASA established a solar flare patrol, and other countries followed suit. The Sun has been under continual observation since then. All of the manned space missions—Gemini, Apollo, Skylab, and Apollo-Soyuz—were scheduled for times when no solar flares were forecast for the mission.

When solar cosmic rays from major solar flares penetrate the skin of a spaceship, they set off secondary rays that must also be absorbed. Space stations in orbit, especially those outside the Van Allen Belts, will have to have protection from space radiation. The International Commission on Radiological Protection set a limit of 0.5 rem a year. The natural radioactivity level in the United States is 0.13 rem a year[7] and the average American receives 0.25 rem a year. A rem is the amount of radiation absorbed by a human being that produces the same effects on human tissues as one roentgen of X rays. Rems are used for measuring the effects of all types of radiation, not just X rays. For example, one rem of neutrons does the same amount of damage as one rem of X rays. Space stations will probably be so shielded that their inhabitants will receive about 0.5 rem a year from space radiation.

Weightlessness and Other Space Effects

Both space radiation and meteoroids are outside perils to spacefarers that can be avoided by shielding or "space weather" forecasts. Weightlessness, however, is inherent in orbiting another object such as a planet, as well as in space away from strong gravitational fields. The zero *g* felt in spacecraft in orbit around the Earth is not the same as weightlessness far from the Earth or planets. A person can appear completely weightless in a spaceship only 320 kilometers (200 miles) above the Earth's surface where the Earth's gravitational field is only a little weaker than it is at sea level. Weightlessness in orbit is simply falling freely. When the Space Shuttle is in orbit, it is falling around the Earth. This phenomenon was predicted three centuries ago by Isaac Newton. Suppose, he said, that a cannon is fired from the summit of a high mountain and the cannon ball lands at its base. Increase the cannon ball's velocity and it will land farther away. Increase it again and it will land still farther away. Keep increasing it and ultimately it will move so far that it will sail around the Earth without landing. In each case the cannon ball is falling; in the last one it is falling around the Earth at the same rate that the Earth's surface curves away from its trajectory. Since everything in the orbiting spacecraft falls at the same velocity and there is nothing to push against, the crew of the ship and everything in it are weightless except for a tiny residual *g*, important only for some delicate experiments.

While weightlessness is very important in space flight, it cannot be duplicated in the laboratory except by dropping objects down a tower. Before manned space flight began, the best that engineers could do was a minute or two in a fast airplane at the top of an arching curve. At the time, some scientists, in predicting the effects of zero *g* on humans, concluded they would be mostly baneful.[8] Some of these predictions were dysbaria (changes in the body due to changes in atmospheric pressure), skin infections, sleepiness, sleeplessness, disorientation, motion sickness, high blood pressure, low blood pressure, need for sedatives, need for stimulants, weight loss, dehydration, bone demineralization, fatigue, muscle atrophy, reduced blood volume, loss of appetite, euphoria, high heart rate, nausea, and pulmonary atelectasis (partial or complete collapse of the lungs).

What have the Mercury, Gemini, Apollo, Skylab, Apollo-Soyuz, Vostok, Voskhod, and Soyuz flights revealed about zero *g*? Humans have now lived in zero *g* for many thousands of hours, and some of the predictions made have turned out to be false. While astronauts and cosmonauts have tended to sleep less in space than they do on Earth, in general sleep has been sound and refreshing. No one has suffered from collapsed lungs or become badly disoriented. In all, zero *g* has not been life threatening.

Some predictions, however, have been borne out. Humans in zero *g* have experienced motion sickness, loss of calcium from bones, muscle atrophy, weight loss, and reduced blood volume in the legs. Blood tends to shift upward to the trunk and head, while fatigue has been a common complaint. There have been cases of sickness in space flight. The Apollo 7 crew contracted colds. In medical parlance, they contracted upper viral respiratory infections. They took prescribed decongestants and soon recovered. One astronaut had a urinary infection and was treated on his return. These were isolated cases that did not recur and proved easy to treat and prevent. Later Apollo crews were kept isolated longer before their missions, and this change in scheduling was effective.

Some of the Apollo astronauts were weaker after their missions. In medical terms, they had experienced a decrease in exercise tolerance. Those who had become weaker regained their former strength in two days after their return. While most astronauts lost weight and had less appetite while in space, this was not serious. On the average, the Apollo astronauts lost six pounds each. Part of the loss seemed to have been caused by dehydration.

Some of these changes seem to be self-limiting, so that they stabilize or later disappear in the flight. Among these are vestibular disturbances—a disorder of the inner ear, which controls our sense of balance; orthostatic intolerance; difficulty in standing upright; and reduction of blood cell mass. Others are not self-limiting, such as shrinking of muscles, especially the calf muscles, reduced capacity for exercise, and loss of minerals from the bones. They will require preventive or remedial treatment for long space flights. Calcium loss has run as high as 2 percent a month, so it could be extremely serious for prolonged missions.

Skylab extended these findings to 84 days, the length of Skylab's longest mission.[9] The longer the astronauts were in zero g, the less time they took to readjust to the Earth environment. The first Skylab crew was in orbit for 28 days and recovered fully in 24 days. The second Skylab crew were in orbit for 59 days, and all recovered within seven days after their return. The third Skylab crew, Skylab 4, were in orbit for 84 days and were back to normal in four to five days.

To the surprise of some scientists, disorientation was rare. The astronaut became oriented simply by making decisions, such as that this partition was a wall and that one a ceiling.

Motion sickness was common in the first day or two, but it went away in a day or so. Often the astronaut could prevent the symptoms of motion sickness by being careful to move his head as little as possible for the first few days in space. Motion sickness depended upon the individual. Significantly, medical scientists could not predict who would be susceptible to motion sickness and who would not be. None of the Skylab 2 crew came down with motion sickness, while all three of Skylab 3 did. Only one of the Skylab 4 crew showed symptoms of motion sickness. Once the astronauts became fully adapted to space, motion sickness disappeared.

Muscle loss continued on the Skylab flights. Blood volume persisted in diminishing in the legs and concentrating in the trunk, but the loss of red blood cell mass seemed to stabilize. Still, calcium loss from the bones went on, slowly but steadily.

Life Science Research in Space

One of the main Space Shuttle programs will investigate the influence of space on humans. The Space Shuttle will carry a Spacelab on many of its missions, and some missions may be completely devoted to medical research. To analyze the functioning of the cardiovascular system, for example, flowmeters will be inserted into the aorta of experimental animals to measure the flow of blood being pumped by the heart. Pressure sensors will measure the flow of blood through arteries and veins. Electrodes will be implanted in the inner ear of laboratory animals to record the tiny currents that control the sense of balance.

Dr. Rufus R. Hessberg, NASA's Deputy Director for Life Sciences,[10] expressed confidence that Space Shuttle flights of from seven to 30 days will not present any serious medical problems. Motion sickness lasts only three to five days and can be satisfactorily treated with anti-motion-sickness drugs; muscle loss will not be appreciable in that time period; and red blood cell loss appears to be self-limiting.

The Space Shuttle will open a new era in space medicine. For the first time, nonastronauts will be members of spacecraft crews. The crew consists of three astronauts, the Commander, the Pilot, and the Mission Specialist, and four nonastronauts, the Payload Specialists. The Commander and the Pilot will fly the Shuttle, the Mission Specialist will manage its operations, and the Payload Specialists will operate the instruments and perform experiments. The latter will have to meet much less stringent medical standards than the astronauts.

The Payload Specialists, who will be scientists and engineers, will be drawn from the general population. As long as the specialist can pass the medical tests, he or she will be accepted regardless of age. Neither will there be a lower age limit. Of course, anyone younger than the early twenties will be very unlikely to have achieved the re-

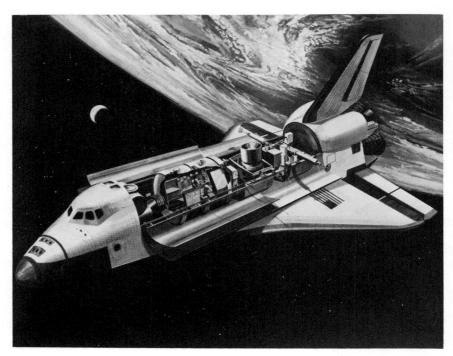

Figure 4-1. Space Shuttle with Spacelab in its payload bay. The drawing shows scientists and payload specialists working in the pressurized forward module of the Spacelab. Aft are instruments mounted on payloads. They are in the vacuum of space. (Courtesy of NASA)

quired scientific education. While there will be no height or weight limits, very tall or heavy people will have difficulty climbing in and out of ports and crawling through the tunnel that connects the Shuttle forward cabin with the cargo bay in which the instruments will be located. Nearsightedness or farsightedness that is compensated by glasses will not be a bar. Payload Specialists will not have to do physical conditioning exercises before a flight but they will be examined both before and afterward and, on their return, will have to be up to their preflight physical status before being released. So, if the reader is in good health, he or she would probably be accepted as physically qualified for Payload Specialist, regardless of age.

What is needed to fully determine the results of space travel is time and research in the space environment. While the Spacelab will go far toward finding the answers, even a 90-day mission in orbit will not be long enough to achieve a thorough understanding.

Spacelab will give American scientists their own facility for investigating the influence of gravity on plants and animals as well as man, all the way from zero g to one g and even higher. This is important, since partial g may be the solution to the harmful effects of zero g. This is an intriguing possibility about which little is known. The astronauts who landed on the Moon were there for a few days at most. A research centrifuge is to be installed in a Spacelab. It will accommodate small animals and plants and will be used to study the effects of one g to zero g. As an example, with this instrument medical scientists could determine the partial g level that is the lower limit for loss of calcium from the bones. The centrifuge may be in operation by 1985.

The most recent American biological experiments were flown on two Russian biosatellites. Cosmos 782 was launched on November 25, 1975, for a flight of 19.5 days,[11] while Cosmos 936 was launched on August 3, 1977, and spent 18.5 days in orbit before coming down.[12] The animals and plants on board were retrieved by Russian scientists, and the American material was returned to the American laboratories. Two experiments were devoted to space radiation, measuring the dose of high-powered nuclear particles that would be absorbed by animals. This would help assess the radiation danger to life in space. The other experiments concentrated on the effects of weightlessness, mostly on laboratory rats. There were also experiments with the wild carrot, Queen Anne's lace (*Daucus carota*), with the eggs of killifish, a kind of minnow, and the fruit fly, *Drosophila melanogaster*.

The Cosmos experiments were carried in a Vostok spacecraft, a hollow sphere 2.4 meters (8 feet) in diameter which had a mass of 2270 kilograms (5000 pounds) and carried a payload of around 900 kilograms (2000 pounds). The atmosphere was maintained near normal, and power was supplied by batteries. The sphere had three levels. On top was a small centrifuge that held rat cages. The centrifuge was 75 centimeters (20 inches) in diameter and ran at 52 revolutions per minute. The American biological material was placed in the one-g and 0.6-g zones of the centrifuge. Beneath the centrifuge, a stationary platform held duplicates of the animals and tissues that were in the centrifuge. Below the platform, 25 rat cages were locked in place.

The experiments showed that weightlessness, at least for the time span of the flights, did not seriously harm the space-borne rats, for they came back clean and healthy. The most obvious effect of weightlessness was shown in their behavior for the first few postflight days. Their sense of balance seemed to have been affected. In one test, falling from a pole, many of the flight rats did not bother to turn as they fell, so they landed tail first, not feet first. This disturbance ended by the tenth postflight day.

There were other revealing findings. Weightlessness accelerated the breakdown of red blood cells, but the animals subjected to one g in flight did as well in this respect as the control animals on the ground.

Ever since the Apollo 11 astronauts reported seeing light flashes, the effect of cosmic rays on the eyes has been worrying the medical scientists. Experiments with cyclotron beams showed that the flashes were due to direct hits by cosmic rays on the retina of the eye. A shower of cosmic rays striking the fovea, the tiny area of sharpest vision in the retina, could badly damage eyesight, while prolonged exposure to cosmic rays in long space flights could cause cataracts. The retinas of the flight rats resembled the retinas of experimental animals exposed to high-energy-particle beams.

In contrast, weightlessness seemed to have little influence on the cells of the wild carrot. Both the centrifuge cultures and stationary cell cultures kept in darkness developed into tiny embryos during the flight. The embryos were transplanted and developed rapidly into normal plants in the nursery. The investigators believe that the cells may have retained an imprint of one g and therefore developed in zero g as they did in one g. Only further experiments can decide the question.

Medicine, Dr. Hessberg remarked, is a "hands on" science. A surgical workbench is being designed for Spacelab. Here, medical scientists will perform research operations on laboratory animals, take blood samples, implant transducers, give injections, take samples of tissues and freeze them for analysis on return to their home laboratories. For example, bone healing in rats in space could be followed by observation of calcium deposition associated with fractures.

Life scientists on board the Space Shuttle will have adequate laboratory facilities. NASA is furnishing life science laboratory equipment that will be specially designed for space use and will be reuseable. There will be animal cages and plant chambers, microtomes, freezers, staining supplies for plant and animal tissues, still and film cameras, microscopes, radiation and acceleration instruments, and even video displays.

The Spacelab I mission will include experiments on plant nutation, the twining of a growing plant. The cause of nutation is believed to be variation of growth rates on different sides of the growing stem tip, due to gravity. This question should be resolved in the Spacelab I flight. The Spacelab II mission will carry experiments on the lignification of plants. Lignin stiffens a plant; it is to a plant what bone is to an animal. Without lignin, a plant could not stand up. The deposition of lignin seems to be gravity oriented. The Spacelab II experiments on lignin will use pine seedlings.

Controlled Ecology Life Support System

NASA has begun a program on life support for future manned spacecraft, CELSS, Controlled Ecology Life Support System.[13] CELSS is a long-range program to develop a biologically assisted life support system to support more people for a longer time than purely physicochemical life support systems such as the Space Shuttle's.

The CELSS program anticipates that the United States will conduct long range, long-duration, or large-crew missions or combinations of all three. CELSSs may be in space stations, lunar bases, or in spaceships on long voyages. If our space flights are to be limited to missions for a few days or weeks, physicochemical systems will do. However, if men are to venture into space for years, or live on the Moon or planets, the cost of supplying them with food will be staggering. A physicochemical life support system for a 100-person, three-year mission that provided food by storage and resupply would have a mass of 95,000 kilograms (208,700 pounds), of which 85 percent would be stored food.[14] A spaceship on such a voyage would be difficult to resupply, unlike a space station in Near-Earth orbit.

In its early years NASA sponsored a modest amount of research on biological life support systems, but in the mid-sixties NASA officials decided that such a system would not be needed for some years so the research was set aside. The Soviets, on the contrary, have been investigating biological life support systems since 1963 and are technically more advanced. They have built a ground demonstrator life support system. It is not an unqualified success, but it does show that they are on the right track.

The first CELSS workshop was held at Ames Research Center on January 9–12, 1979.[15] About 75 scientists and engineers from the academic community as well as from Ames Research Center, Kennedy Space Center, and the Jet Propulsion Laboratory took part. The NASA centers have technical responsibilities for the program and NASA headquarters has overall programmatic supervision. Both the centers and headquarters have let contracts and grants to a number of universities that have the scientists skilled in horticulture, agronomy, ecology, plant physiology, and plant nutrition needed to help develop a hybrid biological-mechanical life support system. Engineers develop the mechanical systems for transporting gases and liquids, for humidity control, temperature control, and environmental control. Dan Popma, the CELSS Project Manager, is himself an engineer.

While physicochemical life support systems can provide air and water, keep the atmosphere fresh, and control temperature and humidity, they cannot supply food. CELSS is investigating the feasibility of growing food from plants that, as part of their life-cycle processes, will also absorb carbon dioxide and generate oxygen. All the wastes will be recycled into plant nutrients. Both biological and physicochemical techniques are under investigation for this purpose. The biological problems may well be the more difficult, because we cannot turn plants off and on by turning a valve or flipping a switch. What is to be done, for example, if the system demands more oxygen or if the food supply is running low? A crop cannot be produced to order in two weeks. Food plants may be planted in small lots at frequent intervals so they will not all reach maturity at the same time. Possibly, CELSS may supply an all-vegetarian diet, or the plant food may be supplemented by eggs from poultry and fish grown in aquaria. An early CELSS might furnish carbohydrates and fats, while animal protein would be kept in storage.

CELSS is divided into four research areas: Human Requirements, Life Support Requirements, Waste Management; and Systems Management, Control, and Ecological Considerations. Human Requirements deals with defining human nutrition requirements; establishing and formulating diet; processing, preparing, and servicing food; preventing toxic effects; habitability and sociological considerations; and "other topics within this general category necessary to insure the proper support of humans such that they can function in an optimal manner for long periods in a closed habitat."[16]

Life Support Requirements include production, supply, and storage of food (both algae and seed plants may be cultivated); nutrition, growth, and harvesting of food plants; secondary food production (feeding plant material to fish in tanks and to poultry or other small food animals); atmosphere revitalization, which includes generating oxygen, removing carbon dioxide, controlling contaminants, and extracting and using water, along with synthesis of food.

Waste Management takes in collecting and treating all wastes, extracting nutrients from wastes, and analysis of wastes for toxic substances. The use of waste to provide plant nutritional elements such as nitrogen will be investigated—for example, extracting nitrogen compounds for use as plant fertilizer.

Systems Management, Control, and Ecological Considerations is concerned with tying all the projects together and making the entire system work smoothly. It includes modeling systems and subsystems, collecting and reducing data, methods of management and control, and ecological considerations that involve system closure, system interface management, and the stability, safety, and reliability of the whole system. System closure means arranging that everything in CELSS is used over and over.

CELSS's first phase will be a ground-based, one-g system that will be as self-contained as possible. Air, food, and wastes will be continually recycled. At first it will not be purely biological. Probably, life support systems will never be purely biological, since they may always have physicochemical backups for safety's sake.

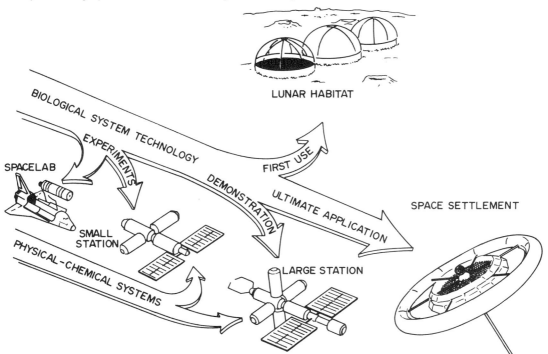

Figure 4-2. Development of life-support systems. Research and development of life-support systems for spaceships, space stations, and lunar bases is a long-range and broad-based program involving biology, chemistry, physics, and technology. (Courtesy of NASA)

Demonstration of a ground-based CELSS will be the first significant milestone, according to Mr. Popma. It will mark the transition from theory to functioning. From the engineering viewpoint, this CELSS demonstrator will be a necessity for verification of proper management and control, one of the most difficult aspects of developing the system. The intent of the project is not to develop a working CELSS on the ground, deactivate it, and then try out a complete system in space. Rather, experiments will be run on the ground and in space. Small life-science experiments can be made in containers stowed in the Space Shuttle's lockers. More elaborate ones can be run in the Spacelab.

Depending upon the progress of space technology, the first biological life support system may be needed by the year 2000. The first fully operational CELSS may be installed in a lunar base. Parts of the system may be in use before then, perhaps in a space station or spaceship that is partly self-sufficient and partly supplied by the Space Shuttle.

The biological life support system will be a simplified model of the grandest of all life support systems, the Earth. CELSS will be a tool that may be of significant help to scientists in understanding the Earth's ecology and in predicting changes in it. CELSS will surely not restrict itself to one or two species of plants. Among other advantages, cultivating a considerable number of plant species will make the system less liable to fluctuating levels of carbon dioxide, oxygen, and so on. It will be "more forgiving."

We can make a safe prediction. Years from now, gardeners will be pursuing a new hobby, "space plants." Gardening enthusiasts will want to have in their gardens, once they are commercially available, the varieties that are growing in the geosynchronous orbit station or at the Lunar Base. A cultivar of a tomato grown at the Lunar Base, say, may be very difficult to tell apart from other cultivars, except by experts—but, no matter, human nature will have its way.

As we have seen, the influence of gravity on plants, animals, and man in all the environments that men will encounter in the solar system can be studied in Earth-Moon space. Zero g occurs in nonrotating space stations, spaceships, and spacelabs in orbit, spaceships traveling with their engines off, both near the Earth and around other planets and their moons. Lunar bases, planetary bases, and natural satellite bases such as bases on Jupiter's moons will be in partial-g environments, except for bases on the small moons such as Phobos and on the asteroids.

Dr. Hessberg believes that for long space flights, partial g could

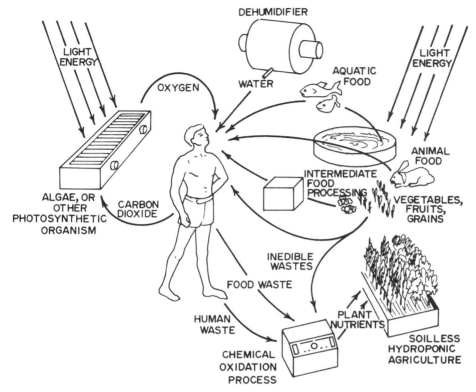

Figure 4-3. Modified ecological life-support system. Life-support systems will imitate, in miniature, the ecology of the Earth. (Courtesy of NASA)

be simulated by a large wheel or torus. Even if it is used only part of the time, the torus might prevent the harmful effects of zero g. By the time that manned voyages to the planets are being actively planned, science should know how to meet the dangers of zero g and how to supply the crew with the necessities of life, no matter how long the mission.

Notes

1. Edward Anders, Dieter Hyman, Michael Lipschutz, and Betty Nielson, "Arizona Meteorite Crater," *American Geophysical Review*, April 20, 1965.
2. Fletcher Watson, "Rockets vs. Meteoroids," *Science*, 104:2696 (1946), 210.

3. N. H. Langton, "Meteors and Spaceflight," in *Spaceflight Today*, ed. Kenneth W. Gatland (London: Iliff Books, 1963), p. 112.

4. Fred L. Whipple, "The Meteoric Risk to Vehicles," *Proceedings of the 8th International Astronautics Congress, Barcelona, 1958*, (New York: Pergamon Press, 1959), pp. 418–428.

5. *Meteoroid Environment Model—Near Earth to Lunar Surface* (NASA SP-8103, March 1969).

6. Shields Warren and Douglas Grahn, "Ionizing Radiation," in *Bioastronautics Data Book* 2d ed. (NASA SP-3006, 1973), pp. 417–451.

7. *Estimates of Ionizing Radiation Dose in the United States 1960–2000*, p. 13, ORP/CSD 72-1, Environmental Protection Agency, Rockville, Md. 20850.

8. Charles A. Berry, "Weightlessness," in *Bioastronautics Data Book*, 2d. ed. (NASA SP-3006, 1973), pp. 349–409.

9. Charles A. Berry, "Medical Legacy of Skylab as of May 9, 1974: The Manned Skylab Missions," *Aviation, Space and Environmental Medicine*, 47:4 (April 1976), 418–424.

10. Dr. Rufus R. Hessberg, Deputy Director for Life Sciences, NASA, personal communication.

11. Susan N. Rosenzweig and Kenneth A. Souza, *Final Report of U. S. Experiments on the Soviet Satellite Cosmos 782*, NASA Technical Memorandum 78525, September 1978.

12. Susan N. Rosenzweig and Kenneth A. Souza, *Final Report of U. S. Experiments on the Soviet Satellite Cosmos 936*, NASA Technical Memorandum 78526, September 1978.

13. Dr. Thora W. Halstead, Life Sciences Division, NASA, personal communication.

14. *The Closed Life Support System* (NASA SP-134, Ames Research Center, April 14–15, 1966).

15. Dan C. Popma, Life Sciences Division, NASA, personal communication.

16. Dr. Thora W. Halstead, Life Sciences Division, NASA, personal communication.

5
PROPULSION: THE KEY TO SOLAR SYSTEM TRAVEL

Propulsion is critical to the space enterprise. All the other elements of space flight are secondary, whether we want to send a satellite into Earth orbit 480 kilometers (300 miles) up or a spaceship to a star many light-years away. Our competence in space propulsion sets the limit on how far we can go in space, how fast we can travel, how much it will cost, and how much we can do. If the United States decided to construct a spaceship to voyage to Proxima Centauri, the nearest star, the vessel could be constructed in ten years which would allow enough time for development of all its systems except one: propulsion. There it would rest at its launching port—imposing, huge, and futile.

What do we want to do in space? Our rocket engines could manage to propel a manned spaceship to Mars, the nearest planet to Earth, but at great expense. Only small, unmanned spacecraft—Pioneers, Mariners, Voyagers, and Vikings—stripped down to their last kilogram (pound) of surplus mass can reach the outer planets. Chemical rockets will suffice only if the United States, Western Europe, and

Japan limit their manned space ventures to Earth-Moon space and the Moon. They are inefficient but, with shrewd engineering, the cost of carrying cargo and men to low Earth orbit, 552 kilometers (345 miles) up, could possibly be brought down to $22 a kilogram ($10 a pound) by the year 2000.

If Americans want to journey into space by the millions each year, tour the Moon and the planets as they now take vacations in Europe; if they want to build lunar and planetary colonies and operate space industrial plants; if they want to send a manned expedition to the stars in the future; in short, if they want to make space travel as cheap and easy as air travel, then the finest rocket engine ever built, the Space Shuttle Main Engine, can only be a beginning. Far more powerful and efficient engines must follow. The United States must undertake a vigorous, long-range space propulsion program to reach these goals.

The propulsion issue is so overwhelming because we live on the surface of a planet that has an escape velocity of 11.2 kilometers (6.95 miles) a second. If a rocket's speed exceeds 11.2 kilometers a second, it may escape from the Earth's gravity, for the Earth's gravitational field will not be able to make it return. This force of gravity decreases in proportion to the square of the distance from the center of the Earth. At 6400 kilometers (4000 miles) altitude, escape velocity is only 2.8 kilometers (9100 feet) per second. This is fine once we get there, but the first 6400 kilometers are the hardest. A rocket needs powerful engines to fight its way through the Earth's gravitational field, power to lift its engines, its propellant tanks, its payload, the entire weight of the rocket. Of course, it must have enough thrust to exceed its weight fully loaded on the launching pad.

How Rocket Engines Work

All rockets function in accordance with Newton's laws of motion. The rocket engine spurts exhaust gas out of its nozzle and so impels the rocket foward in the opposite direction. Rocket engines do not push against anything. They run best in a vacuum and so are well suited to space.

Four properties of rocket engines describe their performance: exhaust velocity, thrust, mass ratio (of the entire rocket, including the engine), and specific impulse. They are sharply different from those, let us say, of an automobile engine. A spaceship may be going at very high speed with its engines stopped. This would not be normal for a car, ship, or airplane, which must continually contend with the friction of the ground, water, or air, and with gravity.

Exhaust velocity is the velocity of the gas or particles moving through the nozzle of the rocket engine, relative to the nozzle. Rocket exhaust need not be a gas. O'Neill's proposed mass-driver rocket engine would hurl rock pellets.[1]

Thrust is measured in newtons in the metric system and in pounds of force in the English system. It is the pushing force developed by the rocket engine. In more exact terms, it is the mass of the propellant ejected per second times the exhaust velocity or speed of ejection. A rocket engine can have a very low thrust with high exhaust velocity, or high thrust with low exhaust velocity, or any combination of the two.

At takeoff, a spaceship standing on its launch pad on Earth must exert greater thrust than its weight. The Apollo 11 spaceship had a mass of 2,902,216 kilograms (6,398,325 pounds) at liftoff. Its five first-stage engines exerted a total thrust of 34.1 million newtons (7,650,000 pounds). If the thrust of its engines exceeds its total weight and its exhaust velocity is greater than Earth escape velocity, 11.2 kilometers (6.95 miles or 36,700 feet) per second, the rocket will do more than take off: it will escape from the Earth. To date, no rocket engine with this exhaust velocity has been built. When it is, it will revolutionize space flight. (The rocket will have to have a mass ratio of 2.187 or more.)

How then did the Apollos reach the Moon and did the Vikings and Pioneers reach Mars and Jupiter? The aerospace engineers get around low exhaust velocity by staging—piling up successively lighter rockets until the top rocket (the last stage) reaches the desired final velocity. With the engines that we have on hand, this is the only way that space vehicles can be sent to the Moon or beyond. Theoretically, stage could be piled upon stage to reach a very high final velocity, but the mighty mountain would yield a very small mouse. The cost and inefficiency would be incredible. This brings us to mass ratio.

Mass ratio is the total mass of the rocket at takeoff divided by the total mass remaining after all the propellants are used up. In other words, it is the ratio of the rocket's "wet weight" to its "dry weight." The mass ratios imposed by the staging of chemical rocket engines plague space-vehicle designers. One of NASA's prime aims is to bring down mass ratios, since too much of a rocket's weight on the pad is propellant. Vanguard I, the first American satellite, had a mass of 13.9 kilograms (30.8 pounds). It was launched by a rocket with a mass of 36,000 kilograms (80,000 pounds), which resulted in a mass ratio of 2600 to 1. This was a great achievement and an engineering abomination.

The best way to reduce mass ratio and the cost of sending payloads into space is to increase exhaust velocity or *specific impulse*, the thrust in pounds due to the flow of one pound of propellant per second. Specific impulse is very important in improving rocket performance, since higher specific impulse means more thrust per kilogram or pound of propellant used. For example, 0.454 kilogram (one pound) of a mixture of hydrogen and oxygen, burning for one second, exerts a thrust of 2031 newtons or 456 pounds of force. Hydrogen-oxygen therefore has a specific impulse of 456 seconds. For specific impulse, the number of pounds of force is equal to the number of seconds. Oxygen-hydrazine has a specific impulse of 368 seconds, while nitric acid-hydrazine, another commonly used combination, has a specific impulse of 315 seconds.

Hydrogen-oxygen has the highest specific impulse of any chemical combination of fuels in use today. In the future, almost all chemical rocket engines will use hydrogen-oxygen. Liquid hydrogen and liquid oxygen are kept near absolute zero. They have been used safely in huge volume for more than two decades, since techniques for liquefying, storing, and handling them have been perfected.

The specific impulse of chemical propellants cannot be raised much beyond that of oxygen-hydrogen, since it is near the limit of chemical reaction energy. A fluorine-hydrogen combination has a specific impulse of 475 seconds, but a fluorine-hydrogen engine could not be used as a first stage. The idea of a rocket ascending through the atmosphere and leaving behind a cloud of hydrofluoric acid vapor would not be accepted by the public. Atomic hydrogen is another possibility, but it does not show much promise of being realized in the near future. Research is being done on compressing hydrogen into a metal under tremendous pressure, but metallic hydrogen is a long way off.

The rockets with which space exploration began had many disadvantages. They were used only once, their mass ratios were high, their payloads were small, and they were largely hand-assembled. The Space Shuttle is phasing out the Delta, Atlas-Centaur, and Titan III-E rockets that have carried spacecraft and satellites into space for two decades.[2]

The Space Shuttle

Each of the five planned Space Shuttles will make over 100 flights in its lifetime. They will carry satellites into space, release them and send them into orbit, retrieve them, overhaul them in space or bring them back to Earth for extensive repairs or modernization. The Shuttles will stay up for as long as 30 days with added life support and power. On some flights they will carry Spacelabs, manned laboratories that will conduct experiments and observations in space.

The shuttle is a hybrid, half rocket and half glider. The Orbiter, a winged rocket plane, is attached to a huge external tank holding propellants. The external tank is 46.8 meters (153.7 feet) long and 8.4 meters (27.6 feet) in diameter. It is divided into a forward oxygen tank and an aft hydrogen tank. Attached on each side of the external tank is a solid rocket booster 45.5 meters (149.1 feet) long and 3.7 meters (12.2 feet) in diameter. These boosters are the largest solid rocket motors to be used in space flight.

At liftoff, the gross mass of the Space Shuttle is 1.99 million kilograms (4.4 million pounds). The solid rocket boosters and the Shuttle's three main engines fire simultaneously with a combined thrust of 29.44 million newtons (6,610,000 pounds), lifting the Orbiter to the altitude of 40 kilometers (25 miles). Their fuel exhausted, the solid rockets drop off and fall away. As they fall, parachutes open up and ease their descent. They are retrieved, brought back to Kennedy or Vandenburg Space Center, cleaned, and refueled for another flight. The Orbiter's main engines carry the ship to its planned orbit, which may be as high as 1200 kilometers (740 miles). As the Orbiter nears its orbit, the external tank is dropped off, its fuel exhausted, and the ship goes into orbit. The external tank is not retrieved. In the standard mode, it reaches 99 percent of orbital velocity just before it is dropped off. In Chapter 6 we discuss possible uses for the external tank.

The three Space Shuttle Main Engines are the key to the Space Shuttle's improved performance. The SSME was designed to be reliable, efficient, easily maintained and repaired, able to start and stop more than 55 times and fire for a total of hours, not minutes as in the past. It must approach the performance of an airplane engine while withstanding the far heavier demands of a rocket engine. It is essential to NASA's plan to reduce the cost of payload in low Earth orbit to a fifth of its present cost.

Table 5-1 compares the SSME with its predecessor, the J-2, the

TABLE 5-1. The J-2 and SSME Engines

	J-2	SSME
OVERALL SIZE	2 m × 2.9 m (6¾ ft × 9½ ft)	2.3 m × 4.3 m (7½ ft × 14 ft)
MASS OF ENGINE	1530 kg (3400 lb)	2835 kg (6300 lb)
EXPANSION RATIO	27.5 : 1	77.5 : 1
LIFETIME	3550 sec (59 min)	27,000 sec (7 hr, 30 min)
FLIGHTS	1	55
THRUST (VACUUM)	1.02 million newtons (230,000 lb)	2.08 million newtons (470,000 lb)
THRUST-WEIGHT RATIO	68 : 1	81 : 1
SPECIFIC IMPULSE	430 sec	455 sec
COMBUSTION-CHAMBER PRESSURE	45,100 newtons/m^2 (760 lb/in.^{2})	176,300 newtons/m^2 (2970 lb/in.^{2})

first large hydrogen-oxygen engine, and reveals the advances the SSME has made. Although larger and heavier than the J-2, the SSME has a higher thrust-to-weight ratio and its expansion ratio is triple that of the J-2. Expansion ratio is the ratio of the rocket nozzle's area at its outer end to that at its inner end or throat, where the hot gases are compressed as they flare out of the combustion chamber. Expansion increases the speed of the exhaust gases. The SSME's higher specific impulse is also proof of its more efficient use of propellants and is almost the maximum possible for hydrogen-oxygen. The increased chamber pressure also aids efficiency by allowing high expansion at low altitude. The J-2 engine was designed to last for only one flight, while the SSME is expected to last for 55 flights before replacement.

The Orbiter is equipped with three SSME engines. If J-2 engines had been installed in their place, six J-2s would have had to have been squeezed into the space taken by three SSMEs and would have furnished less total thrust. The J-2s would have had to have been replaced after each flight. The SSME marks the transition from throwaway rockets to spaceships.

Cost per kilogram of payload is the critical issue in space enterprise. The lower the cost, the more can be done in space. Table 5-2 shows the steep drop in cost per kilogram (or pound) in orbit in the first two decades of the Space Era.[3]

The size of the payload affects the cost per kilogram or pound. The first American satellite was so expensive because it was so small. The payloads of the expendable rockets were in the tens of kilograms, usually less than 140 kilograms (300 pounds), until the manned space program began. Mercury, Gemini, and Apollo each set records for mass of payload. The first American manned flight into orbit placed 1315 kilograms (2900 pounds) in orbit on February 20, 1962. On its fifth test flight, Saturn IB, the predecessor of the moon rocket, lifted 17,100 kilograms (37,700 pounds) into orbit. On November 9, 1962, Saturn V on its first launch took 42,506 kilograms (93,709 pounds) into orbit.

The Space Shuttle will be the main vehicle of the Space Transportation System, the world's first spaceline. NASA engineers are confident that they can cut operating costs as they gain experience with its operation. They have consulted officials of Trans-World Airlines, one of the largest airlines in the world, to gain the benefit of the airline's experience in planning the operations of the STS.

Beverly Henry and Charles Eldred of the NASA Langley Research Center have analyzed the problem of reducing the costs of space transportation and have come to these conclusions.[4] Reuseability is mandatory, and complete reuseability should be the goal. Through experience, the cost of running the STS may be reduced to a tenth of that of the first flights. Keeping the Space Shuttles fully utilized and in constant operation will help. Cutting the turnaround time from two weeks to two days would be highly beneficial, since any vehicle, be it airplane, taxi, barge, tanker, or space shuttle, earns money only when it is in motion.

The Space Shuttle will markedly bring down the cost of launching satellites. For example, a communications satellite, Syncom IV, proposed by Hughes Aircraft Company to be launched into geosynchronous orbit, 35,900 kilometers (22,300 miles) high, would be released by the Space Shuttle at an altitude of 296 kilometers (185 miles). The satellite, which would be designed to fit into the shuttle's cargo bay, would carry its own rocket motor. After being lifted out of the cargo bay and released, the satellite would spin slowly; then, when at a safe distance, its motor would be fired, sending the satellite up to geosynchronous orbit. Launching the Syncom IV into geosynchronous orbit with Delta would cost $13.5 million, with Atlas-Centaur, $24.5

TABLE 5-2. Trends in Payload Cost* in Low Earth Orbit[a]

Date	Cost per Kilogram	Cost per Pound	Vehicle
Jan. 31, 1958	$220,460	$100,000	Juno I
1965-1971	$7463	$3385	Delta
1971-1976	$5225	$2370	Atlas-Centaur
1980—est.	$617	$270	Space Shuttle

[a]Low Earth orbit = 552 kilometers or 345 statute miles
*Figures are given in 1975 dollars. Table reflects transportation costs per pound of payload in orbit. Many other factors also affect payload cost.

million; and with Titan III, $58.3 million. Space Shuttle would do it for $5.3 million. It is estimated that the Space Shuttle will save over $11 billion between 1980 and 1991 by replacing expendable rockets.

Initially, the Space Shuttle will transport a 29,500-kilogram (65,000-pound) payload to a 185-kilometer (115-mile) circular orbit. With reduced payload, it could reach an altitude of 1200 kilometers (740 miles). For sending payloads to geosynchronous orbit, the Space Shuttle could carry either an Interim Upper Stage or a Spinning Solid Stage. These spacecraft could be loaded into the cargo bay.

Geosynchronous orbits are very useful. A satellite in geosynchronous orbit makes a complete orbit around the Earth in the same time that the Earth rotates on its axis. If placed over the Equator, the 24-hour orbit keeps the satellite hovering over one spot on Earth so it is, relative to the Earth, fixed in space and admirably suited for communications and weather satellites. The satellite is 35,890 kilometers (22,300 miles) above the Earth's surface.

The Interim Upper Stage is a solid-propellant rocket that can carry up to 2270 kilograms (5000 pounds) to geosynchronous orbit. It will be a cheap vehicle for taking satellites to geosynchronous orbit. For smaller loads destined for geosynchronous orbit, the smaller Spinning Solid Stage will be used, a solid-motor rocket that spins in flight.

Once the Space Transportation System is fully operative, it will make about 60 flights a year. For the period 1980 to 1992, 560 flights are planned. Space flight will become routine. By 1990, hundreds of men and women from all over the world will have flown in space.

Advanced Space Shuttles

The future of space enterprise depends largely upon the volume of cargo carried into space. The five Space Shuttles together will be able to carry 1.8 million kilograms (4 million pounds) each year to low Earth orbit. If the demand for space traffic increases to 45 million kilograms a year (100 million pounds), then spaceships much larger than the Space Shuttle will be needed. But what would require transporting 45 million kilograms of cargo each year into space? Only a huge multipurpose space facility, enormous space structures such as a system of giant communications satellites, or a lunar base or all three could generate such volume.

As space traffic grows, the Interim Upper Stage will be replaced by a Manned Orbital Transfer Vehicle based at the Low Earth Orbit Space Station. It will always remain in space, ferrying people to work in geosynchronous orbit. Eventually, the MOTV will be used to construct a geosynchronous orbit space station.

Improved engines will be needed for these tasks. The SSME will probably be followed by the Dual Expander Engine that has been advocated by two aerospace engineers, Robert J. Salkeld of System Development Corporation and Rudi Beichel of Aerojet Liquid Rocket Company.[5] It combines the advantages of burning a dense hydrocarbon fuel at launch with the high specific impulse of hydrogen-oxygen in space. The Dual Expander Engine would use three propellants: liquid oxygen, liquid hydrogen, and RP-1 or some other suitable hydrocarbon fuel.

The Dual Expander Engine will be two engines in one. Instead of a single combustion chamber, it will have a central chamber surrounded by a concentric, ring-shaped chamber with both chambers ejecting into the same nozzle. Liquid oxygen and RP-1 are pumped into the central chamber at takeoff and burn at the pressure of 41.2 million newtons per square meter (6000 pounds per square inch). Oxygen and hydrogen burn in a preburner, driving the fuel-pump turbines, and then roar into the annular chamber at the pressure of 21.1 million newtons per square meter (3000 pounds per square inch). On reaching the vacuum of space, the engine switches to burning only hydrogen and oxygen.

The Dual Expander Engine marks a significant advance over the SSME. While hydrocarbon-oxygen has a lower specific impulse than hydrogen-oxygen, hydrocarbon fuels are much denser than hydrogen for equivalent energy output and so require much smaller fuel tanks. At takeoff both combustion chambers fire to generate a high thrust, while at high altitude only the annular chamber is firing; its lower thrust maintains the acceleration at a low g, to keep the crew comfortable. The engine is efficient and compact. A Dual Expander Engine with a thrust of 3.01 million newtons (692,700 pounds) in space would have a mass of 1958 kilograms (4350 pounds) and a thrust-to-engine-weight ratio of 156:1 compared to SSME's mass of 2835 kilograms (6300 pounds) and ratio of 83:1. Its small size—its nozzle is only 2.1 meters (7 feet) long and wide—would suit it admirably for the SSTO, the logical successor to the Space Shuttle.

The SSTO (Single-Stage-To-Orbit) vehicle would be a winged, streamlined spaceship looking like a super-Orbiter. It would be a "one-piece" spaceship. Nothing would be discarded in flight. Several aerospace firms, notably the Boeing Company and the Martin-Marietta Corporation, have prepared designs for an SSTO.

The Boeing spaceship would take off horizontally, carried on a sled that would speed up to 658 kilometers (409 miles) an hour to help send the ship aloft with its engines going full blast. The spaceship, however, could not ascend on its own, and its flights would be limited to specially equipped space centers. In contrast, the Martin-Marietta ship would take off vertically and land horizontally like the Space Shuttle. It would be 61 meters (200 feet) long with a 61-meter wingspan and a mass of 1,925,000 kilograms (4,235,000 pounds) fully loaded and 203,000 kilograms (446,000 pounds) empty. NASA engineers have predicted that SSTOs in this class, carrying about 27,000 kilograms (60,000 pounds) per flight, would reduce costs to low Earth orbit to as low as $33 a kilogram ($15 a pound). For comparison, airlines flying from New York to Los Angeles have a payload cost of about $1.60 a kilogram ($0.73 a pound).

For very heavy loads a huge freight spaceship, the Heavy Lift Launch Vehicle, is being considered.[6] In one version, the aft fuselage would be changed to a separable entry vehicle for the crew. On a due-east flight, after main engine cutoff, the entry vehicle would separate and fly a ballistic couse down to the Australian desert, using a parachute and retrorockets to land. With four reuseable liquid rocket boosters, the HLLV could carry 152,000 kilograms (335,000 pounds) to low Earth orbit for a payload coat of $20 a kilogram ($9 a pound). Operating costs could be lowered even more by having the HLLV unmanned and guided to its destination by radio command. Even so, by the year 2010 aerospace engineering will have reached the limits of the chemical rocket engine.

Harry O. Ruppe,[7] a German aerospace engineer, is pessimistic on the prospects for reducing space transportation costs by as much as the NASA engineers predict. He believes that it may be brought down to $687 a kilogram ($308 a pound). LEO, in Ruppe's opinion, can only be the beginning of space utilization. The next step, transporting cargo to geosychronous orbit, would cost about $4100 a kilogram ($1880 a pound), while lunar orbit might cost $6200 a kilogram ($3800 a pound). And that, according to Ruppe, does not include the cost of transporting the cargo to the lunar surface in a lunar shuttle. Only actual experience will determine whose set of predictions are correct, if either.

The Space Shuttle and even its successors, the SSTO and the HLLV, will still be only coastal spaceships venturing at best to near-space islands such as the Earth orbiting space stations and no farther than the Moon. Nuclear engines will be needed to radically slash the expense of space flight and to send manned expeditions out into the solar system. Since chemical reactions have an insufficient energy maximum, nuclear energy must replace chemical energy. It is the hope for true solar system rocket engines.

The Nuclear Rocket Program

A rocket runs most efficiently when its exhaust velocity equals its forward speed. Exhaust velocity is proportional to the square root of the temperature of the gas divided by its molecular weight. Since hydrogen is the lightest gas, it has been selected as the propellant for nuclear rocket engines. To reach a velocity of 320 kilometers (200 miles) a second, which would be reasonable for solar system spaceships, the exhaust would have to have a velocity of 320 kilometers a second, a specific impulse of 32,800 seconds, and a temperature of 4.1 million degrees Kelvin.

The energy that can be liberated in nuclear reactions is far greater than that of chemical reactions. In fission, up to 0.12 percent of atomic mass is converted into energy. This would give a maximum possible specific impulse of 1.5 million seconds! Fission's potential energy far exceeds the requirements for a 320-kilometer (200-mile)-a-second spaceship. Fission, though, presents difficult engineering problems. Fissioning atoms do not just break up into convenient atomic fragments for rocket thrust. Fission reactions also give off gamma rays, neutrons, and high-speed electrons that heat up the reactor and make it radioactive. Converting this energy into a form usable for space propulsion is a major problem for engine designers.

Research on nuclear space propulsion began a few years before the United States began its space program. The Rover Project to develop a nuclear rocket for manned space flight was established by the Atomic Energy Commission and the Air Force on November 2, 1955, over two years before the first American satellite was launched on January 31, 1958. A lengthy series of tests and models was planned, for a whole new field of technology had to be mastered. The nuclear reactor fuel was to be uranium 235 in graphite fuel elements, and the propellant would be liquid hydrogen. With its low molecular weight, hydrogen would easily reach high exhaust velocity and, therefore, high specific impulse. Rover was to undertake space exploration not possible with all-chemical rockets.

On March 6, 1957, a site in the Nevada desert was authorized. About 90 miles northwest of Las Vegas, it covers 140 square miles. On February 16, 1962, it was designated the Nuclear Rocket Development Station, the center where Rover reactors, engines, and stages would be tested.

The design of nuclear engines is governed by their high temperatures and high power densities. The hotter the nuclear reactor, the more energy available for space propulsion. One design for a nuclear rocket engine simply substitutes a small nuclear reactor for the combustion chamber. Instead of burning two gases such as oxygen and hydrogen, hydrogen flows through the hot reactor where it is heated to a high temperature and spurts out of the rocket nozzle. This is the basic concept of the solid-core nuclear engine.

Not even structural metals with high melting points can withstand the blasting heat of a nuclear furnace without being cooled. Regenerative cooling is as necessary to the nuclear solid-core engine as it is to the chemical rocket engine. The hydrogen flows through tubes in the reactor wall and absorbs heat before passing into the reactor chamber. Even with regenerative cooling, only refractories with very high melting points can be used, such as hafnium carbide, niobium carbide, zirconium carbide, tungsten (3380°C), and carbon (3500°C). But some of them strongly absorb neutrons instead of reflecting them back to the reactor chamber. The neutrons were to be used in the reactor and, hopefully, would not be absorbed by the reactor and engine. Only those refractories with high melting points that did not strongly absorb neutrons were selected, such as zirconium carbide and graphite. Graphite was the choice but, unfortunately, it reacts chemically with hot hydrogen. The solution to this problem was to coat graphite with niobium or zirconium carbide.

Soon after NASA was established, the new agency reached an agreement with the Atomic Energy Commission to jointly conduct the nuclear rocket program. On August 31, 1960, a joint office, the Space Nuclear Propulsion Office, was set up. The Rover program had to be extensive. The Los Alamos Scientific Laboratory would get assistance from other agencies and companies. The NASA Lewis Research Center was given responsibility for development of nonnuclear engine components, while NASA Marshall Space Flight Center would work on advanced vehicle technology. NASA laboratories would do research on gaseous reactors, the next step beyond solid-core reactors. They would be more efficient than the solid-core reactors, since they would run at higher temperatures.

Early in the program, the endpoint of the solid-core engine program was planned to be the Reactor-In-Flight-Test (RIFT). The nuclear engine was to be called NERVA—Nuclear Engine for Rocket Vehicle Application. The first NERVA would be a flightworthy nuclear engine. A nuclear-powered third stage would replace the chemical S–IVB stage of a Saturn rocket. The first and second chemical stages would be retained. The reactor in NERVA would go critical, or be "turned on" to begin fission, by radio command when the stage reached an altitude of 1600 kilometers (1000 miles). Six RIFT stages would be tested on the ground at the Nuclear Rocket Development Station in Nevada. The first flight test was planned for 1968, to be launched from the Kennedy Space Center.

The first experimental nuclear rocket reactor, Kiwi A, was tested at the Nuclear Rocket Development Station on July 1, 1959. As its weight was greater than its thrust, it was named after the kiwi, a flightless bird of New Zealand. The purpose of Kiwi A was simply to learn how to build and test a nuclear rocket reactor. The reactor was transported several kilometers by rail from its assembly building to its test cell. The test, which consisted of running the reactor at full power long enough for it to undergo a realistic cycle of operations, was a complete success. Kiwi A consumed only three kilograms (6.6 pounds) of liquid hydrogen a second and generated only 100 megawatts (5000 pounds of thrust).

While the reactor was being developed, other related projects were going on. In July 1961 Aerojet General Corporation was awarded a contract for the design and the construction of NERVA. NERVA would be derived from the design of the Kiwi B, the second-generation reactor, which did best in the tests. In May 1962 Marshall Space Flight Center contracted with the Lockheed Missile and Space Company to design rocket stages that would serve as test flight vehicles for a nuclear rocket engine. Westinghouse Electric Corporation, the leading subcontractor to Aerojet General on NERVA, would construct a series of these Kiwi-based reactors.

Kiwi B would have a newly designed reactor and new pumps, use 30 kilograms (66 pounds) of liquid hydrogen a second, and generate 1000 megawatts (50,000 pounds of thrust). Two large storage dewars, each holding 15,000 kilograms (33,000 pounds) of liquid hydrogen, had to be built to provide enough hydrogen for the test runs. In December 1961, the first Kiwi B, running at low power, was successfully

Figure 5-1. NERVA, Experimental Reactor A-1. This was one of a series of experimental nuclear reactors of the NERVA program that was to lead to a nuclear rocket engine. (Courtesy of NASA)

tested. The critical test came in September 1962 when it was run on liquid hydrogen. The mechanical parts (the nozzle and the turbopumps) performed well, but the reactor spouted white-hot chunks from the nozzle. On November 30, 1962, Kiwi B was again tested. The reactor ran smoothly, yet flashes of light shot out of the nozzle.

When the reactor was disassembled by remote control, the engineers found that most of the fuel elements had broken up. Kiwi B had to be redesigned. The modified Kiwi B was tested in August 1965 and this time it ran smoothly.

In late 1965, the first Westinghouse reactor, NRX-A2, based upon Kiwi B4, was tested. In its subsequent engine system test, NRX-A6, in which the main engine components were tested, the reactor ran for 60 minutes. This was in December 1967. The test of the complete experimental engine, NERVA-XE, was also successful. Building an engine for flight testing was to come next.

This step was never taken, for NASA was forced to reduce its budget. First, plans for a 990,000-newtons (200,000-pounds) thrust engine had to be reduced to one with only 333,000 newtons (75,000 pounds). In the summer of 1968 NASA's budget for the next fiscal year was slashed savagely. It was reduced by $362 million; 1600 employees were to be fired and 2000 contract employees to be dropped. Construction of four Saturn rockets were canceled, exploration of the Moon after Apollo 17 was halted, and the nuclear rocket engine project was weakened.

Ironically, despite the cut in funds, the prospects for NERVA's success were growing brighter. It had only to be allowed a fair trial to prove its usefulness. By the spring of 1969, more than a dozen reactors and engines had been tested. The NERVA-XE, the latest of the series, was tested 28 times for a total of 3.5 hours that spring and summer. Reactor technology reached 10 hours and the specific impulse had been raised to 850 seconds, almost twice that of today's SSME engine. Then, in February 1970, the nuclear engine project was again cut back. In the FY 1971 NASA budget, NERVA was assigned only $38 million with an equal sum from AEC. This was trivial, for from the start of the project about $1.1 billion had been spent on it, and an equivalent sum was needed to bring it to flight status.

The Space Shuttle was now coming along as the successor to Apollo as NASA's chief project, and NASA officials did not want to endanger its prospects with Congress. The end of the Apollo Project was approaching, and NASA officials hoped that the Space Shuttle would open a new era in space. To save the Rover program and to keep its options for the future open, it was to be reoriented to be tied into the Space Shuttle system. The Space Shuttle would carry the NERVA engine, tanks, and payload into low Earth orbit, where the components would be assembled to power a nuclear spaceship that could carry heavy payloads from low Earth orbit or geosynchronous orbit to lunar orbit or into the solar system.

President Nixon then sent to Congress a Fiscal Year 1972 budget that led Senator Clinton Anderson, Chairman of the Senate Committee on Aeronautical and Space Sciences, to exclaim,

> *I know of no advanced technology program which has been successful than the nuclear rocket propulsion program and I do not understand the recommendations of the fiscal year 1972 budget submission to reduce*

the level of effort to a mere holding action. In fact, I would have difficulty understanding any reduction in this important program.[8]

Nixon had recommended $30 million while AEC and NASA had asked for $110 million to prepare for the first flight test, which had now been set for 1978. But the President prevailed, and hundreds of people working on the nuclear rocket project were fired. The Nuclear Rocket Development Station was closed down and left to caretakers. In January 1973 the Atomic Energy Commission and NASA ended the development of nuclear rockets.

Dr. James J. Kramer, Associate Administrator of NASA's Office of Aeronautics and Space Technology, stated: "In the 15 years of nuclear rocket research and development, an outstanding history of technical achievement was recorded and the capabilities of nuclear rocket propulsion were thoroughly demonstrated."[9] Demonstrated, but not tried. After 15 years of hard work by thousands of engineers, scientists, technicians, and skilled workers, the great project was abandoned.

Harold B. Finger, manager of the joint AEC-NASA Space Nuclear Propulsion Office,[10] said that the termination of the Rover Project in 1972–1973 was due to contraction of NASA's missions. Nuclear propulsion had been intended for very-heavy-payload, very-high-energy, very-long-range missions to Mars and Venus. When plans for manned interplanetary voyages were dropped, the need for nuclear power in space disappeared. However, Finger added, should NASA resume planning for manned exploration of the planets, nuclear-powered spaceships would be required for these missions.

The solid-core nuclear engine could have been the first of a succession of advanced-technology engines, each generation reaching higher specific impulse and exhaust velocity to culminate in the fusion engine that would take mankind to the brink of interstellar flight.

The limitations of solid-core engines as well as their potential were recognized even before research was started on them. The barrier to continuing improvement of the performance of solid-core engines is the melting point of materials. The hotter the reactor, the higher specific impulse it produces. The solid-core reactor must be hotter than its propellant gas, for heat flows only from hot to cold regions. The highest-melting-point substance is hafnium carbide, which melts at 4150°C (7500°F). The maximum specific impulse of solid-core engines is about 1000 seconds.

This obstacle could overcome by shifting from solid to gaseous fission. Gaseous reactor engines could possibly reach a specific impulse

Figure 5-2. An artist's concept of a nuclear-powered spaceship. The hydrogen tanks help shield the crew in the forward compartment of the spaceship from any radiation given off by the reactor. (Courtesy of NASA)

of 20,000 seconds. Both radiators and regenerative cooling would be used to keep the engine from melting. Further, in a gaseous-core engine, the fissioning gas fireball is surrounded by the propellant gas, which flows around it and out the nozzle. The propellant carries most of the heat with it. In the solid-core engine, the solid fissioning uranium is inside fuel elements, while the propellant gas flows around them and then out through the nozzle. The heat sources surround the outflowing propellant gas.

Research began on gaseous-core engines about the same time that work started on the NERVA project. Since the gaseous-core engines would operate at higher temperatures and pressures than the solid-core engines, gaseous core research was confined to experiments and analytical studies. Two kinds of gaseous-core engines were investigated: the coaxial-flow reactor and the nuclear-light-bulb reactor. The NASA Lewis Research Center headed the coaxial-flow reactor project, while United Aircraft Research Laboratories studied the nuclear-light-bulb reactor under contract to the AEC-NASA Space Nuclear Propulsion Office.

According to George McLafferty of United Aircraft, the project director, the coaxial reactor engine might look like this:[11] A heavy sphere 3.6 meters (12 feet) across contains 50 kilograms (110 pounds) of hot fissioning uranium-233 gas. The radiation of this fireball has a power of 22,000 megawatts. Liquid hydrogen is pumped in and flows through tubes inside the inner surface of the sphere. It turns into gas and flows inside the interior cavity, directly absorbing the heat radiating from the fireball, and rushes out of the nozzle. The reactor heats 100 kilograms (222 pounds) of hydrogen a second to the specific impulse of 1800 seconds, giving it the thrust of 17,800,000 newtons.

The pressure in the chamber would be 1000 atmospheres—about 100 million newtons per square meter (15,000 pounds per square inch). The reactor would have a mass of 127,000 kilograms (281,000 pounds) including a moderator shell 0.76 meter (2.5 feet) thick inside the pressure shell. The moderator, made of beryllium and heavy water, would keep the fission reaction in the uranium-235 going by reflecting back the neutrons it produced. Otherwise, the reaction would die out as the neutrons escaped.

One of the critical problems in designing an efficient coaxial engine is to keep the uranium gas from escaping through the nozzle with the hydrogen. One solution is to keep the heavy uranium gas moving slowly while the light hydrogen speeds around it and out the nozzle. Even so, uranium would escape at about 1 percent of the hydrogen flow rate. This would require feeding in uranium to keep the reaction going.

The nuclear-light-bulb engine completely prevents this loss. In it, the uranium gas is enclosed in a transparent shell, and heat transfer is entirely by radiation. Since the shell is comparatively cool, it does not melt as the radiation passes through it. The radiation is absorbed by the propellant hydrogen gas flowing around the "light bulb."

This design has seven concentrically arranged cavities. Neon gas flows just inside the transparent silica wall to cool it. Fused silica is about the only material that will do the job, but the gamma rays and neutrons given off by the fission reaction can make it opaque. However, when heated to 800°C (1470°F), it once again becomes transparent to the desired wavelengths. The transparent wall would therefore be maintained at 800°C, at which temperature it would absorb only 1 percent of the radiation from the fireball.

The gaseous nuclear engine program was divided into four

phases. Phase I would determine by analytical studies whether either type could function if its high-temperature problems could be solved. When it ended in 1967, the result for both engines was yes. Phase II was a series of tests using radiofrequency and arc heaters to test materials and design ideas at reactor temperatures. It went well. Phase III was to be a series of small-scale nuclear tests, in which miniaturized engines would be run by a current of neutrons from a "driver" reactor that would make their uranium fission. Phase III was to take two to six years. Phase IV was to be the construction and test of a full-scale nuclear engine. If NASA had not been starved for funds and the Atomic Energy Commission had been allowed to continue its research and development programs, in all probability a gaseous-core nuclear rocket engine would be well on its way to flight status. Such an engine would also serve as a prototype for an advanced-technology stationary site nuclear fission electric generating plant.

NASA had only two gaseous nuclear reactor projects left—after Project Rover ended—at Los Alamos and at the Langley Research Center. Their purpose was to study the physics of gaseous-core reactors, their safety and control. An experimental reactor was set up at Los Alamos in December 1976. The engineers cannibalized parts such as beryllium reflectors from the nuclear reactors at the Nuclear Rocket Development Station. They planned to achieve a temperature of more than 1000°C (1800°F) for a series of tests that could eventually lead to practical gaseous nuclear reactor engines. But NASA was not granted funds to continue the gaseous-core reactor work and, like the Rover Project, it had to be terminated at the end of fiscal year 1978.

Over two decades have passed since nuclear engine development began, yet the United States still does not have a nuclear rocket engine. Without it, deep-space exploration is blocked. This is not a scientific failure or an engineering failure. It is a political failure of the first magnitude.

Notes

1. Gerard K. O'Neill, *The High Frontier* (New York: William Morrow & Co., Inc., 1977), p. 138.
2. Myron S. Malkin, "The Space Shuttle—Key Considerations," in *Proceedings of the XXV International Astronautical Federation Con-*

gress, Amsterdam, September 30–October 5, 1974 (Elmsford, N.Y.: Pergamon Press, Inc., 1975), pp. 241–259.

3. Financial Manager, Expendable Launch Vehicles Office, Office of Space Flight, and History Office, NASA, personal communications.

4. Beverly Z. Henry and Charles H. Eldred, "Advanced Technology and Future Earth-to-Orbit Transportations Systems," *Third Princeton/AIAA Conference on Space Manufacturing Facilities, Princeton, N.J., May 9–12, 1977*, 77–530, American Institute of Aeronautics and Astronautics.

5. Rudi Beichel, "The Dual Expander Engine," *Astronautics and Aeronautics*, November 1977, pp. 44–51.

6. M. W. J. Bell, "Advanced Launch Vehicle Systems and Technology," *Spaceflight*, 29:4 (April 1978), 135–143.

7. Harry O. Ruppe, "Limits of Space Transportation," *Journal of the British Interplanetary Society*, 30 (1977), 76–78.

8. *Aviation Week and Space Technology*, March 1, 1971, p. 14.

9. *NASA Authorization for Fiscal Year 1978*, Part 3, p. 1515, 95th Congress, First Session, Senate Committee on Commerce, Science and Transportation, Subcommittee on Science, Technology, and Space.

10. Harold B. Finger, General Manager, Center for Energy Systems, General Electric Company, personal communication.

11. George H. McLafferty, "Gas-Core Nuclear Rocket Engine Technology Status," *Journal of Spacecraft and Rockets*, 7:12 (December 1970), pp. 1391–1396.

6
COLONIZING EARTH-MOON SPACE

The region from low Earth orbit, 555 kilometers (345 miles), to the Moon, 400,000 kilometers (250,000 miles) or so away, is the testing ground for the space enterprise. Here, the techniques for living in space, efficiently moving men and cargoes, providing space power, and using space industrially must be worked out. When they are, the solar system will be open to the human race.

Earth-Moon space is small on the solar system scale and readily accessible. The first space voyagers, the Apollo astronauts, took four days to reach the Moon, averaging about 4000 kilometers (2500 miles) an hour. This is a very slow speed for space. The lunar flight via Space Shuttle to Earth Orbit Space Station and via Lunar Spaceship to Lunar Orbit Space Station and thence to the Moon via Lunar Module to the Moon should do better. The flight to low Earth orbit should not take more than a few hours, and flight from lunar orbit to the lunar surface should be even quicker. If the Lunar Spaceship averages only 6.4 kilometers (4 miles) a second, the trip from space station to space station will take about 18 hours and from Earth to

Moon only one day. Such a flight would be readily acceptable to travelers. The flight from Washington, D.C., to Sydney, Australia, takes 24 hours, yet the round trip is popular with tourists. The lunar spaceships would be larger and more commodious than today's airliners and the Kennedy-Lunar Base flight should not lack customers.

The idea that time, not distance, separates places, has revolutionized civilization. When space travel becomes common, people will think no more of traveling a few hundred thousand kilometers in space than they now do of traveling a thousand kilometers on Earth, as long as they can do it in the same time. Once they can get to the Moon in a day, such a trip will seem as routine as a trip to Australia does today.

The success of the space enterprise is usually estimated in terms of Earth-received benefits, such as so many billions a year from weather forecasting, from satellite telephone relays, from earth resources observations, and so on. A better criterion might be the number of people who go into space each year. From the ten astronauts a year of the Apollo era to the 300 or 400 a year of the Space Shuttle era is a big step forward, but only the second one. Suppose that the Space Transportation System is extended to the Moon, as we have indicated, and a regular Earth-Moon schedule is begun with a weekly round trip by a 100-passenger lunar spaceship. Then 5000 travelers will visit the Moon a year, a traffic an order of magnitude beyond that of the Space Shuttle but still miniscule compared to long-distance Earth travel. Yet even at this level, space travel will begin to become part of our vacation plans.

NASA Plans for Earth-Moon Space

What are NASA's plans to use Earth-Moon space? NASA officials are deeply concerned with their agency's future in space. Planning is a critical NASA activity, since the agency must plot its course in the vast space ocean. But plans remain dreams on paper until approval is received from the Office of Management and Budget and from Congress. Since the end of the Apollo Project in 1972, NASA has been the target of politicians who regard space as a cost, not as an investment, and has been kept on a restrictive budget.

The Space Shuttle is the first large-scale NASA project that will pay for itself and it will make acute the question of what direction the space agency will take. What should be the next advance beyond the

Space Shuttle? Space now offers to American business the prospect of billions of dollars a year from space industry. Should science and exploration take a back seat to industry and profit?

In June 1974 Dr. James C. Fletcher, then Administrator of NASA, appointed a study group to consider the future of the United States in space. It was charged with developing a list of desirable civilian space activities for the period 1980 to 2000, to define the research and development required for potential commercial uses of space, and to identify benefits flowing from space exploration and research. The purpose of the study was to identify future space objectives, not to develop detailed mission or program plans. The group was made up of 20 NASA engineers and scientists and one representative of the Air Force. These 21 were the nucleus. More than 180 NASA experts contributed their ideas and opinions, and people from many occupations—such as writers, economists, astronomers, and physicists—were consulted. Men and women from many major universities, including several in Europe, federal agencies and science-oriented companies, took part. All this was done while the participants carried on their usual duties. This study, "Outlook for Space,"[1] represents a consensus of the American space community. Let us consider its conclusions:

1. A major increase in emphasis and in resources should be directed toward Earth-oriented space programs.
2. Among humanity's needs that are particularly amenable to the use of space-oriented data are monitoring and prediction of climate and severe weather, prediction of crop production and water availability, and monitoring of changes to the environment.
3. The increasing shortage of available energy warrants an intensive investigation of the technical, economic, and environmental feasibility of space power generation and hazardous waste disposal in space.
4. The transition between the research and development of space systems for Earth-oriented needs and the operational use of these systems is critical. Planning for this transition should be an integral part of the earlier phase and will require substantial resources.
5. The multiple use of operational, remote-sensing spacecraft will be necessary to economically exploit the utility of the nation's Earth-oriented space program. The development and management of these spacecraft will require broadened and innovative cooperative arrangements between government agencies at the federal, state, and local levels as well as between government and the private sector.
6. The space science program of the future should continue exploratory re-

search but should increasingly be focused on specific objectives selected on the basis of their expected contribution to fundamental scientific questions.

7. Scientific questions of particular current significance deal with the beginning of the Universe; the nature of black holes and quasars; gravity; the evolution of the planets and their atmospheres; the nature of the Earth's climate; and the search for extraterrestrial life, including intelligent life.

8. Social, economic, and technical forces will one day cause humans to further explore and exploit the solar system.

9. A major increase in attention and resources should be directed toward data management and information extraction.

10. There should be increased emphasis on theory, development of predictive modeling, and data analysis in support of future space missions.

11. The United States needs to maintain its international leadership in space communications and, therefore, should increase its efforts to advance communications technology and to develop and demonstrate new uses.

12. An active program should be pursued to exploit the zero-gravity environment of space to increase the understanding of material science and basic physical, chemical, and biological processes.

13. The United States should develop a permanent Space Station to fully exploit the zero-gravity environment of space for basic research, to develop a full understanding of the ability of humans to live and work in space for extended periods, and to provide a potentially useful instrument of U.S. foreign policy.

Conclusions 1 to 5 are little more than approval of NASA's current activities: NASA should continue its efforts to make satellites serve human needs for weather prediction, communications, crop survey and prediction, water supply, pollution monitoring, and so on. Conclusion 3, however, indicates that NASA should consider solar-power satellites.

Conclusions 6 to 8 deal with science and exploration. One wonders why the investigation of the evolution of the planets and their atmospheres by unmanned spacecraft is more important than the manned exploration of the bodies of the solar system, which can yield much more knowledge of them in the long run. With the masses of information accumulated, the question of planetary evolution will be derivative as it is for the Earth. The aim of solar system exploration should be to learn as much about the rest of the solar system as we know about the Earth. Planetary probes are only the beginning, which makes one ask why "social, economic and technical forces will one day cause humans to further explore and exploit the solar system." That day is here.

The last five conclusions are unexceptional. Conclusion 12 is really part of conclusion 13—that the United States should develop a permanent space station. This is the next step beyond the Space Shuttle and has the broad support of the space community. NASA hopes to have a space station by 1984. Why wait?

Space Colonies and Solar Power Satellites

In this situation, reawakening public enthusiasm for space adventure together with the caution of NASA officials, who have had to deal with congressional antagonism for over a decade, have left an opening for new leadership. Dr. Gerard K. O'Neill of Princeton University is advocating a huge space colonization project in near-Earth space.[2] His idea has made considerable progress toward public acceptance. The space colonies would be vast hollow cylinders. Each colony would consist of two cylinders connected at their ends, counterrotating to create an artifical gravity. They would be inside-out worlds complete with air, water, landscaping, villages, farms, factories, day-and-night cycles, sunlight, and even seasons.

The specifications for even the smallest space colony, Model One, are staggering. Model One would be a kilometer (3280 feet) long and 200 meters (656 feet) in diameter, revolve once every 21 seconds, house 10,000 people and be completed by 1988. Its total mass would be about 520,000 metric tons, including 20,000 tons for aluminum for structure, 10,000 tons of glass, 50,000 tons of water, a 1000-ton generator plant, 2800 tons of initial structure, special hardware, machines, and tools, 420,000 tons of soil, rock, and construction materials, 5400 tons of liquid hydrogen, 2000 tons allowance for people and their possessions, and 500 tons of dehydrated foods.

The space colony would not be located in low Earth orbit 552 kilometers (345 miles) up or in geosynchronous orbit 35,900 kilometers (22,300 miles) above the Earth. It might be situated in deep space 386,000 kilometers (240,000 miles) away at L-4 or L-5, orbiting slowly around either of these Lagrangian points and so remaining in a fairly stable position relative to the Earth and the Moon. An alternate location would be a two-day-duration orbit at an altitude of 60,700 kilometers (37,700 miles). In any orbit the space colony would be extravagantly expensive to construct.

Space transportation would be the overriding factor in constructing this artificial satellite. The bulk of its material would be mined on the Moon and fired to L-5, or its lower target, by an electromagnetic accelerator called a mass driver. The cost, from the Moon to L-5, is estimated at $30.1 billion. O'Neill estimated the cost of lifting the required equipment and material from the Earth to L-5 at $935 a kilogram ($425 a pound) for a total of $8.5 billion in 1972 dollars. Transporting the work crew from the Earth to L-5 would cost $22 billion, while the project, as originally planned, would take six years and cost a total of $51 billion.

Nor is this all. The space colony project would depend heavily upon new, untried technology that is not even on the drawing board, especially for the critical transportation factor, even though O'Neill says the plan uses present-day technology. The Space Shuttle has a maximum cargo capacity of 29,500 kilograms (65,000 pounds), but that is only for a due-east circular orbit at an altitude of 185 kilometers (115 miles).[3] The Space Shuttle could reach an altitude of 1200 kilometers (800 miles), but only by trading payload for extra propellants. In this case it could carry only about two tons of payload. The Space Shuttle would have to make 34 flights to lift the 10,000 metric tons (2,200,000 pounds) up to only 185 kilometers. If the Interim Upper Stage with its 2300-kilogram (5000-pound) capacity were used to take the cargo from 185 kilometers to geosynchronous orbit, 35,900 kilometers (22,300 miles) up, it would have to make 440 flights and still be short of its target. A new space transport would be needed, perhaps the Heavy Lift Launch Vehicle. The job cannot be reasonably done with the present Space Transportation System.

O'Neill's romantic scheme is a chain of linked, exuberant assumptions, each costing many billions and requiring development of new technology. The space colony supporters justify it by claiming that it will provide the United States and the world with a new, cheap, inexhaustible kind of energy—electricity from sunlight in space. O'Neill's space colonies would be financed in time by solar-power satellites, enormous structures in geosynchronous orbit that would convert the energy of sunlight into microwaves that would be beamed to huge Earth-based antennas called rectennas. Here, the microwaves would be converted to electricity to be fed into the country's power network.

Why undertake this gigantic project that could cost hundreds of billions of dollars? O'Neill contends that it would solve some of the major problems facing humanity:

1. Bringing every human being up to a living standard now enjoyed only by the most fortunate.
2. Protecting the biosphere from damage caused by transportation and industrial pollution.
3. Finding high-quality living space for a world population that is doubling every 35 years.
4. Finding clean, practical energy sources.
5. Preventing overload of Earth's heat balance.

These are admirable goals. Can they be met by space colonies and solar power satellites for less cost and more effectively than by other means such as nuclear energy? This is the critical question. All space plans should be measured by the same criteria:

1. Comparative cost: Can it be done for less by another technique?
2. Comparative effectiveness: Are its aims better achieved by other means?
3. Comparative disturbance of the environment: Does it disturb the natural environment more than other projects with the same aims?

For many years, comparative cost and comparative effectiveness have been the principal grounds for making planning decisions. Only recently has effect on the environment been taken into account.

O'Neill's space colonies will become reality only if the solar-power satellite is found to be superior to nuclear energy. A small experimental solar-power satellite could settle this question, but it would have to be placed in geosynchronous orbit for a realistic test. With the Space Shuttle only able to take it into low Earth orbit, one would have to provide a means of increasing its altitude.

The standard of living is fundamentally the standard of energy consumption. For example, in 1976 the United States consumed 345 million BTUs (British Thermal Units) per capita, while Egypt used only 14 million BTUs per capita.[4] Since the beginning of civilization, every rise in living standards has been due to an increase in available energy. The domestication of draught animals, the wheel, the sailing ship, the water mill, the windmill, all raised human wealth above what could be secured by human muscle. When steam was introduced in the eighteenth century, per capita income started a steep rise stimulated later on by the internal combustion engine, electricity, and, in more recent times, nuclear power. If poverty is to be abolished in the backward countries, their energy supply must be enormously expanded by modern technology.

Petroleum has raised per capita income in the Arab oil-producing countries such as Saudi Arabia and Kuwait but only at the expense of the Western oil-consuming countries. It has been a huge transfer of wealth, not a creation of wealth. The United States, Europe, and Japan have lost hundreds of billions to OPEC, the Arab-controlled oil monopoly, and consequently suffer inflation and unemployment. OPEC's steady pushing upward of the world price of petroleum has especially harmed the poor countries. They could hardly afford to pay $3.14 a barrel in 1970, and still less over $30 a barrel in 1980. The only reason that nations with big oil fields are able to victimize nations without them is that petroleum is not universally distributed. The lesson of petroleum is that the poor nations need a type of energy that is freely available to all.

Solar space power would be even more susceptible to monopoly than oil, for at present only the United States and the Soviet Union are capable of placing large solar-power satellites in geosynchronous orbit. Possibly the European Space Agency, Japan, and China could develop this capability by greatly expanding their space programs. In any case, a network of solar-power satellites would require an investment of hundreds of billions over a period of decades and a huge supporting space technology infrastructure. The poorer nations would have to depend upon the superpowers, or, at least, the advanced nations, for their solar electricity, either as purchasers or as junior partners. Would they be content to do so? Their complaints about such things as Western communications satellites in geosynchronous orbit suggest they might not be.

Solar-power satellites would introduce a brand-new kind of pollution—the bombardment of the atmosphere with microwaves. Pro-SPSers estimate that 60 solar-power satellites could furnish 10 percent of the electricity demand of the United States.[5] The transmitting antenna of the solar power satellite would be about 1.5 kilometers (5000 feet) across, while the receiving antenna or rectenna would be about 10 by 14 kilometers (6 by 8.5 miles). Consequently, each SPS would radiate a huge microwave beam over two kilometers (a mile) across when it hit the top of the atmosphere, spreading to an area of about 125 square kilometers (50 square miles) as it reached the ground. If solar-power satellites generated all of the nation's electricity, about 64,000 square kilometers (25,000 square miles) of the atmosphere over the United States would be radiated with microwaves, not for a few minutes but steadily year in and year out. What would their effect be on people? What would be their effect on the climate? On radio and television? No one knows. Certainly radioastronomy would be paralyzed. Radiotelescopes might have to be removed to space or a lunar base. Optical astronomy would be badly affected. For decades, astronomers have decided upon new sites for telescopes only after prolonged searches for the locations with the best seeing. They would be upset by a new band of bright spots curving across the night sky. The celestial equator would be all too conspicuous! Observations near them would be useless. Would the world's great observatories have to be abandoned for space astronomy?

Providing room for the world's population, which is doubling every 35 years, is utterly futile. We cite no data—only appeal to the reader's common sense. Sooner or later, and preferably as soon as possible, the world's population increase must be halted. Otherwise, life on this planet will become unbearable for sheer lack of space. This is beyond dispute. Even if the whole solar system is settled, a limit to land area will be reached. Nor is the goal of a stable world population visionary. The advanced nations have come close to stationary population levels, and the rest of the world will follow as they adopt birth control measures. Contraceptives are easily billions of times more cost effective than settling hundreds of millions of people in huge space colonies. The battle for a livable world will be won in the bedroom, not in outer space.

In the past three decades, climatologists have become deeply concerned over the possibility that human activities are raising the Earth's average temperature. There are two chief possibilities. The heat given off by power plants, vehicles, factories, buildings, and the rest of the apparatus of civilization may be a major cause. The enormous volume of carbon dioxide given off by burning gasoline, coal, gas, and oil may be another. Analysis has led to general agreement that man-made heat will not significantly influence the temperature of the atmosphere on a global basis for the next century or so. Carbon dioxide appears to be the culprit. If carbon dioxide continues to accumulate in the atmosphere at its present rate, the average air temperature on Earth may rise by 8°C (14°F) by the year 2200, while the temperature at the North and South Poles will rise by 24°C (43°F).[6] The warming of the Antarctic ice cap could melt a large region of it, and the consequent rise in sea level might be rapid. Coastal cities all over the world would be submerged.

Even by the rosiest projections, solar-power satellites would merely replace coal-fired, oil-fired, and nuclear power plants as sources of electricity for the United States. They would not affect the major source of air pollution, gasoline-burning motor vehicles, which emit the overwhelming mass of air pollutants. For example, and this is typical, 96 percent of the carbon monoxide and 79 percent of the hydrocarbons in the air in the metropolitan Washington area in 1976 were due to motor-vehicle exhaust. They poured 2400 tons of air pollutants into the air over the area every day.[7] Carbon monoxide is the product of incomplete combustion. The exhaust of a 100 percent efficient gasoline engine would be carbon dioxide and water vapor.

Solar-power satellites, even if they generated the entire electricity supply of the United States, would therefore not appreciably reduce air pollution since they would not remove its cause.

The choice between solar-power satellites and nuclear power plants centers on cost per kilowatt of electricity generated. Nuclear power plants are designed and constructed by established technology on the earth. The design of nuclear reactors could be standardized and they could be mass produced. Standardization of nuclear power plants could creat sufficient demand to warrant mass production for domestic and overseas purchasing. This would help reduce the huge deficit in the American balance of payments, the principal cause of inflation. In fact, the Soviet Union is building a factory to make nuclear reactors and plans to export them.

Solar power satellites must be lifted to geosynchronous orbit, 37,000 kilometers (22,300 miles) above the Earth, at gigantic expense. Peter Glaser, Vice-President of Arthur D. Little, Inc., the inventor of the solar-power satellite, envisages an SPS delivering 5000 megawatts to its rectenna far below. The satellite would have a mass of 18.2 million kilograms (40 million pounds) and with its solar collector panels would be 4.93 kilometers by 13.1 kilometers (3.05 miles wide by 8.1 miles long).[5] The gigantic structure would be assembled in two stages by sending material up to low Earth orbit (552 kilometers or 345 miles) for preliminary assembling and then up to geosynchronous orbit for final assembly. Space Shuttles would be inadequate. Only a specially designed spaceship such as the Heavy Lift Launch Vehicle could undertake it. The HLLV would have to make up to 70 flights to GSO at an estimated cost (in 1975 dollars) of $40 to $120 a kilogram ($18 to $54 a pound).

Glaser estimated that the 5000-megawatt SPS would cost $7.6 billion or $1500 a kilowatt, furnishing electricity at the busbar (delivered to the electrical power network) for 2.7 cents a kilowatthour. If the HLLV did not lower the transportation cost to about $45 a kilogram ($20 a pound), the cost per kilowatthour would be much higher. The Boeing Company's study of a solar-power satellite estimated its design would cost $13 billion and cost 5 cents per kilowatthour.[8]

In October 1977 the United States had an electrical capacity of 550,000 megawatts. To maintain this capacity, 110 solar-power satellites would have to be emplaced at an estimated cost of $836 billion and in all probability the cost would run far higher.

In contrast, nuclear energy is the cheapest way to generate electricity. According to the Atomic Industrial Forum, in 1975 nuclear electricity cost 1.5 cents per kilowatt hour.[9] This cost could be reduced if the time taken for the construction of nuclear plants were decreased from the current ten to fifteen years. In contrast, Japan builds nuclear plants in four years. If the design of reactors were standardized and new reactors built on existing sites, such plants could be built in two to three years. The creation of an Energy Production Agency, which would set the same safety standards for all forms of energy production, would conclusively demonstrate the relative safety of nuclear energy to other forms for fixed sites.

Safety is always relative. There has never been a fatality in an American nuclear plant that was nuclear-related. A study by J. P. McBride and his associates at Oak Ridge National Laboratory showed that an advanced coal-fired power plant can emit as much radioactive effluents as a nuclear plant.[10] The Allen steam plant at Oak Ridge burns 675 million kilograms (743,000 tons) of coal a year and releases 6.7 kilograms (15 pounds) of uranium and 13.4 kilograms (30 pounds) of thorium into the atmosphere each year. In both cases, the radioactive dose per person is slight, well below the standards set in the Code of Federal Regulations of 25 millirems per year for the total uranium fuel cycle. The toxicity of the hundreds of thousands of metric tons of ash from the coal power plant is non-nuclear, yet a major environmental problem.

According to Dr. Bernard L. Cohen of the University of Pittsburgh, disposal of radioactive waste is not a major problem.[11,12] The Federal Government must reach a decision on a storage site and proceed. If the spent fuel from nuclear reactors is concentrated in solid form in glass rods, the rods could be sheathed in steel cylinders and stored in a deep mine. There they would stay as their radioactivity slowly and steadily decayed. Even if all the electricity in the United States were generated by nuclear energy, the containers for the next thousand years would take up an area 30 kilometers (12 miles) square and 3 meters (10 feet) high if they were placed 10 meters (30 feet) apart.

There are some far greater non-radioactive dangers in our daily lives. About 10 billion kilograms (10 million tons) of chlorine gas are produced each year in the United States, enough to kill 400 trillion people. At its average rate of production, the world population could be wiped out in less than an hour. In truth, release of chlorine by railroad and truck accidents is so unusual that such events are reported immediately by television, radio, and the newspapers across the country even though deaths due to chlorine inhalation are infrequent.

Higher temperature nuclear reactors could be designed to generate hydrogen by dissociating water above 2000°C. The steam would break up into hydrogen and oxygen, both valuable chemicals. Hydrogen makes an admirable automobile fuel. It is the cleanest of all fuels and produces water when properly burned.[13] In short, replacing gasoline with hydrogen would end air pollution, for all practical purposes. Existing automobile engines could be easily modified to use hydrogen so the American automobile industry would not have to spend billions in retooling to fabricate a brand-new type of engine.

There are two drawbacks: storage and cost. Hydrogen could be carried in cylinders at high pressure, but they would be very heavy and hard to handle. Hydrogen could be liquified, but cars would then have to have cryogenic systems. The best way would be to store the hydrogen chemically combined as a spongy metal, called a hydride, in a metal tank. Several firms and institutions are working on developing a lightweight storage container that will hold enough hydrogen to power a car between normal "fill-ups". The hydrogen is released by warming the container with the hot exhaust.[14]

The cost of hydrogen must be reduced to the equivalent cost of gasoline on a mile-traveled basis to be competitive. While a gallon of hydrogen has a mass of 0.3 kilograms (0.6 pounds), it is the most energetic of fuels and yields 2.7 times as many BTU as an equivalent amount of gasoline.[15] On an energy-equivalent basis, hydrogen requires four times the volume of gasoline. However, the difference in the price of hydrogen and of gasoline is not very great, only about twice as much on an energy-equivalent basis in mid-1979. Mass generation of hydrogen might, in time, reduce its price well below that of gasoline.

The progress of nuclear technology in the United States has been held back by its opponents. The reactors now in operation could be replaced by advanced, high temperature reactors that would generate electricity directly from nuclear reactions, instead of using atomic energy to heat water to heat steam to drive turbines to drive generators to generate electricity. The capital cost and consequently the price per kilowatthour would be reduced.

Fusion, now within two decades of commercialization,[16] would devastate the prospects for solar-power satellites. It has none of even the alleged disadvantages of nuclear power and can be set up anywhere on Earth or, for that matter, in the solar system. Space fusion reactors would not require square kilometers of collector panels. They would be compact compared to solar-power satellites of equivalent power output.

The Space Station

What can be done is not necessarily what should be done. The full value of the Space Shuttle cannot be realized as long as it merely ventures briefly into nearby space and soon returns. It should have a destination. A space station could be assembled in low Earth orbit at a very early date. NASA has been conducting or underwriting space station studies for two decades and does not lack for ideas.[17] There is no need to start by building a grandiose "multipurpose space facility." Such a project would cost billions, take time to generate a substantial revenue, and arouse much opposition. It would be wiser to start small and expand gradually. As has been suggested, insofar as possible it should be made up of available components such as the Space Shuttle External Tank. It is critical that it be placed high enough in orbit to stay up permanently—that is, at about 640 kilometers (400 miles) altitude, below the Inner Van Allen Belt. The space station, in time, would serve several purposes, many of them offering the prospect of financial returns:

1. Zero-*g* experiment station: It can serve as a laboratory for research on the effects of zero gravity on humans, animals, plants, and equipment for long periods, extending to years. This is the critical research facility of the space enterprise. (See Chapter 4.)
2. Closed-ecology life support system: Completely self-sufficient life support

systems are essential not only for space stations but for lunar and planetary bases and colonies and for long space voyages. This will be a continuing research and development project. The problems are intricate, and there will always be room for improvements.

3. Space research medical center: Will space benefit patients with such diseases as arteriosclerosis? If this is demonstrated, the space station may include a space hospital.

4. Space observatory: Skylab underscored the value of continual observation of the Sun from space by observers. A permanent solar observatory would in itself justify the cost of a space station. There could also be a stellar and planetary observatory.

5. Industrial laboratory/pilot plant/production plant: An industrial laboratory for investigating industrial processes in space would lead to a pilot plant for experimental production, which, in turn, could lead to commercial production for shipment back to Earth.

6. Satellite facility: Minor repairs and maintenance of satellites in orbit can be done with the Space Shuttle. Once facilities are available in a space station, there will be no need to take satellites down to Earth for major repairs or renovation. In time, the Satellite Facility could design, fabricate, and launch satellites to order.

7. Spaceport: Keeping a fleet of spaceships permanently in space would save fuel that would otherwise be consumed in flying them back to Earth and then launching them back into orbit. Why waste all that good kinetic energy? With the space station as the first space stop, it would be a transfer point for travelers to geosynchronous orbit, the Moon, and, later on, the planets. As a spaceport, the space station could cut the cost of space transportation. Liquid oxygen and hydrogen would have to be brought up from Earth. But only the propellants would have to be ferried, not the whole dead weight of the spaceships. If ice were discovered in the permanently shadowed areas around the north and south poles of the Moon, a lunar base could dissociate the ice and ship liquid oxygen and hydrogen to the space station at low cost. This would give the lunar base a strong economic justification. Together with a lunar-orbit space station, it would sharply reduce the cost of space travel.

8. Spaceship construction base: As the satellite facility and the spaceport expanded their services, they would take on the construction of unmanned spacecraft and possibly even of manned spaceships.

Putting a space station together from the Space Shuttle's External Tanks is the only way to construct one for less than a billion dollars. The External Tanks would, in a way, cost nothing, since otherwise they would be destroyed. A space station assembled from specially designed and constructed components would cost at least $2 billion and probably more. In addition, since the Space Transportation System

Figure 6-1. An artist's concept of a space station under construction. This picture shows a space shuttle, a space taxi, and construction methods used. (Courtesy of Rockwell International, Space Systems Group)

will have at least two Spacelabs and possibly five, one for each Space Shuttle, the European Space Agency would most likely be willing to build more of them to be incorporated in the space station.

Marshall Space Flight Center's plan for making a space station from External Tanks is straightforward.[18] Instead of dropping off its External Tank when it reaches 99 percent of orbital velocity, the Space Shuttle leaves the huge cylinder—46.86 meters (153.68 feet) long and 8.38 meters (27.58 feet) in diameter—in orbit. The Shuttle carries only a 15,800-kilogram (35,000-pound) payload, since the forward liquid oxygen tank of the ET has been divided by a bulkhead to provide 56 cubic meters (2000 cubic feet) of living space for the three-person crew. This space has been equipped with electrical, life support, and communication systems. A docking hatch has been fastened to the ET's nose cap, and air and water are stored in the section between the oxygen and hydrogen tanks. On its next flight, the Shuttle brings up the space station crew, a solar-power module, and a docking module with an airlock.

Figure 6-2. Europe, seen from space. This photo was taken on May 29, 1978, by the Meteostat Satellite of the European Space Agency. It shows a view of the Earth as it might be seen from a space station. (Courtesy of the European Space Agency)

The Shuttle docks with the External Tank, the solar panels are attached and connected, the ET's oxygen tank is purged of all traces of propellants and then is pressurized. The crew enters its new quarters and makes them comfortable. The oxygen tank could be used as a laboratory. The cavernous hydrogen tank, 30 meters (about 100 feet) long, could be made over into the first space industrial laboratory or pilot plant. It could even serve as a hangar for spacecraft.

Once the basic space station is operating smoothly, adding External Tanks and Spacelabs could be routine. The combination of the capacious space of the external tank with the sophisticated instrumentation and equipment of the Spacelab would make expansion of the capabilities of the space station very easy.

The Space Shuttle can remain in orbit normally for only 30 days, while the space station would be in orbit permanently. The space station crew members might stay there in 60-day shifts, but they could remain for 180 days or even longer. Rotating the crews at long intervals would reduce operating costs and help long-duration research projects. Its much greater size, diversity of facilities, living space, and, in time, growing number of inhabitants would make the station far more appealing for a long stay than the Space Shuttle.

As the space station developed, it might progress toward self-sufficiency. A space power reactor would eliminate the need for regularly bringing up fuel from Earth as well as dispense with huge solar panels that could impede observing. CELSS (closed-ecology life support system) research would furnish food, water, and a continually regenerating atmosphere. Once pilot-plant manufacture of a product was proven, space could be rented for commercial production and "export." The space station would begin to be self-sustaining as well as self-sufficient.

The space station would be a source of spectacular copy for the world's news media. Journalists would be eager to visit it and report on its diversified activities, at least until it became taken for granted.

Space manufacturing has been treated as a possible bonanza for industry that could yield billions of dollars a year in revenue. A word of caution is apropos. Until research on the physical and chemical processes that underlie production proves the advantages of space manufacturing relative to the manufacture of the same products on Earth, the future of space manufacturing will be in doubt. Its products may be negligible—or conceivably they may some day exceed, say, the gross national product of the United States for 1980.

The National Research Council sponsored a study by an ad hoc committee of scientists and engineers of the processing of materials in space.[19] Their conclusions were cautious:

> *The Committee has not discovered any examples of economically justifiable processes for producing materials in space and recommends that this area of materials technology not be emphasized in NASA's technology program. The Committee has identified some activities in which experiments in space and on earth can be expected to contribute usefully to the understanding of materials processes or to the preparation of specialized exemplary materials. Research and development along these lines seems appropriate, with the initiative resting with the prospective investigators.*

The Committee's caution may be excessive, but the potential of space manufacturing cannot be determined until extensive basic research is done in space, followed by pilot-plant production in space.

Space Communication Systems

Space industrialization is not synonymous with space manufacturing, for it also includes applied space projects. Communications is well established in space and its future is assured. More than a billion dollars have been invested in communications satellites, and the demand for space communications is growing year by year. The region for communication satellites is the circular belt 294,000 kilometers (165,000 miles) long that lies 35,900 kilometers (22,300 miles) above the Earth's surface, the geosynchronous orbit. GSO is now populated by 50 communications satellites and in five years will be populated by a hundred or more, doing over $2 billion worth of business.[20] They are serving radio and television networks, telephone companies, computers, and defense communications—and that is only the beginning.

Ivan Bekey of NASA has argued very persuasively for placing in geosynchronous orbit big, powerful satellites, so powerful and sensitive that millions of people could use them with rooftop antennas a meter (a yard) across.[21] The more advanced the comsat—the more powerful and sensitive—the less power the customer on the surface needs in order to employ it. Bekey and his associates have done much planning for electronic mail, widespread educational television, and individual radiotelephones that would make "Dick Tracy wristwatch radio" a reality.[22] Other intriguing possibilities include voting/polling wrist sets, personal navigation wrist sets, emergency/rescue wrist beacons, disaster communications wrist radios, all aircraft traffic control, burglar intrusion detectors, vehicle and package locators, nuclear materials locators, library data sharing, forest fire detectors, fault movement detectors (to warn of coming earthquakes), anticollision radar, interplanetary television links, and much more.

Space communications started in 1962 with weak satellites and huge ground-based antennas. Place the power source in space, and the cost per user will be very small. Suppose, Bekey suggested, an advanced comsat for the wrist radio telephone network were built. It would cost about $300 million. The wrist radios, however, would cost only $10 each, a price that the 2.5 million users could willingly afford. If a cheaper, less powerful satellite costing only $50 million were substituted, each user terminal would cost over $1000. We leave the decision to the reader.

One proposed system that would be universally popular would be advanced television broadcasting. A satellite in geosynchronous orbit weighing only 6300 kilograms (14,000 pounds) and broadcasting with

50 kilowatts power could cover the United States and enable the viewers to watch television broadcasts from any television station in the country. The viewer would only have to buy a special one-meter (three-foot) wire dish antenna and have it installed on his roof, or the dish could be incorporated in the set by the manufacturer.

Let us carry this idea one step further. Why not form a system of international comsats that would enable the viewers to tune in on any television station in the world? Besides the 988 American stations, there are 25,000 foreign stations. What a wealth of choice!

The benefits would be far reaching, and there are no known harmful side-effects. Americans know far less about current events in the rest of the world than they should for their own good. By watching television broadcasts from Europe, Asia, Africa, Australia and Latin America, they would see for themselves what is happening over the Earth. Conversely, the Asiatics, Europeans, Africans, and Latin Americans would witness the happenings in our own country. The world would begin to become a neighborhood. When the Apollo 11 astronauts landed on the Moon, 600 million people watched on television. This is an indication of what space communications can do, and must do, to bring about a world community.

Of the 150 or so countries in the world, most are controlled by dictatorships, each with its own Iron Curtain. No matter what their political wavelength, all dictators vehemently oppose free access to news, for dictatorships are sustained by censorship. Just the same, these governments would be under pressure to cooperate, for if they tried to jam the comsat television transmissions (not an easy matter), their television stations would be barred from the network. Television signals can dissolve Iron Curtains.

The giant communication satellites would be mutliantenna platforms, radiating hundreds of beams, receiving thousands of signals simultaneously, and communicating with each other as well as with earth transmitters. Each platform would have a power generator, which would be a nuclear reactor, and other common facilities for running many diverse communications and sensing systems. The satellite's operations would be controlled by a central computer operating near the speed of light.

It would be sensible to keep these expensive machines operating for as long as possible. The early satellites had a designated lifetime of one or two years, while the latest will run for seven years. These limits are due mainly to the satellite's consuming the fuel with which its jets keep it in position. They counteract the gravitational forces of the

Earth, Moon, and Sun that slowly draw it away from its assigned position. If astronauts could visit the comsats to refill their fuel tanks, they could last for decades, if not for centuries. While there, the astronauts could also replace defective parts, run tests, and change the equipment. Comsats could be updated as required, and they would become permanent installations in space.

A geosynchronous space station could at least partly pay for itself by servicing the GSO (geosynchronous orbit) comsat network. The astronauts would have to have their own spacecraft, based at the space station, to visit the satellites. It would be economical to run, since it would not have to fight the Earth's gravitational field. Traveling at the same altitude and at almost the same orbital velocity as at launch, it would use little fuel on its service missions.

The Space Transportation System would have to be extended to support the GSO space station. The Space Shuttle cannot reach geosynchronous orbit, so a new space vehicle would have to be built, the Orbital Transfer Vehicle. The Orbital Transfer Vehicle would shuttle back and forth between the two space stations, bringing supplies and transferring people from one to the other. The Orbital Transfer Vehicle would most likely be fueled at the LEO (low Earth orbit) space station. Since its propellant would have to be lifted from Earth, the OTV would be more efficient if it were nuclear powered, for a nuclear OTV would have higher specific impulse, carry more cargo per kilogram (pound) of propellant, and require only one propellant instead of the two used by a chemically powered OTV.

A nuclear OTV could easily become the first lunar spaceship, carrying modules to lunar orbit, where they would be joined together to make the first lunar orbit space station. From this station one-stage lunar shuttles, resembling the Apollo Lunar Modules only much larger, would descend to the Moon's surface to set up the first lunar base.

At first the lunar base could be a crude affair, much like the first bases set up in Antarctica. Small though it might be, however, it would begin an advance from which there would be no turning back. The ultimate goal would be to build a lunar civilization. The immediate aim would be much less ambitious—learning how to live on the Moon, doing research, and exploring the Moon kilometer by kilometer.

The Moon will be an exceedingly interesting and challenging place to live. The unearthly, beautiful terrain will tempt the adventurous and offer outdoor adventure—climbing craters, descending rilles,

Figure 6-3. The Moon, from lunar orbit. This is an oblique view of the far side of the Moon, taken by Apollo 17 in lunar orbit, looking southward. The large crater is I.A.U. 308, which is 50 miles across. (Courtesy of NASA)

exploring the unknown beyond the near lunar horizon in spacesuits. Life on the Moon need not suffer from dullness.

The first industrial venture of the lunar base could be the extraction of oxygen and glass from the regolith (lunar soil), since the lunar rocks are about 41 percent oxygen and about 20 percent silicon.[23] Extraction of oxygen in large volume could help make the base self-sustaining financially. Liquefied oxygen from the Moon could make large enclosures feasible, cut the cost of propellant for the lunar shuttles and for the lunar spaceship if the latter had chemical engines, and supply the Earth space stations with liquid oxygen for their spacecraft. It would reduce the cost of space transportation to the Moon,

Figure 6-4. A future lunar base. This is an artist's concept of a future lunar base near the south pole of the Moon. The astronauts' lander is on the right and a power station is in the background. (Courtesy of NASA)

making it cheaper to ferry equipment, supplies, and people to build up the lunar base.

The second product to receive early attention would be glass. Plagioclase, a very common lunar mineral, could be melted to make an excellent glass. Heat would be needed, so a small nuclear power plant would be one of the first shipments to the young lunar base. While useful as a supplement, solar energy would be available only half the time. Lunar nights are two weeks long, and the base would need a constant energy supply.

A third product would be hydrogen. If ice is discovered at the poles of the Moon, the lunar base will have made a discovery comparable to finding a 500-billion-barrel oil field on Earth. Water and hydrogen then will not have to be imported from Earth. If not, hydrogen may be wrung from the solar wind, either by heating the top regolith to extract the traces of solar hydrogen deposited by the solar wind or by stretching an ion scoop, a vast electrified wire net, many square kilometers in area, to draw the solar wind into a scoop or funnel and extract the hydrogen directly with magnetic fields. However, the solar wind is so thin that the scoop would have to be many kilometers across to extract more than a few kilograms of solar hydrogen a year.[24] With cheap nuclear energy on hand to extract elements from the regolith, even the slightest trace of hydrogen would be saved.

Lunar industry will be inorganic. With carbon, nitrogen, and possibly hydrogen having to be imported, synthetic substitutes for plastics, paper, fabrics, lumber, and numerous other organic materials will be sought by the lunar base chemists. Even if every element in the lunar base is carefully recycled without significant loss, hydrogen will be critical for the expansion of the colony. The supply of hydrogen may well be the lunar colonists' most severe problem. It will be far more valuable than gold or platinum, for it will be used to synthesize water with oxygen from lunar rocks, and as fuel for spaceships and lunar rovers.

A Lunar Base Tour

Let us imagine what a trip to the Moon will be like by the year 2010—which, after all, is not very far off.

In 2010 Kennedy Space Center remains the main American spaceport and is much larger and more elaborate than it was in the early Space Shuttle era. It is as busy as a large airport. Spaceships are taking off and landing several times each day. We are in the passenger lounge waiting to board. We have had our medical checkups and all have passed, which is not hard to do unless you have a cold or other infectious disease, for NASA is not anxious to export germs into space. Our baggage, including hand baggage, is being carefully weighed, for the total mass of the shuttle must be determined to a kilogram. After the preliminaries are completed (you do not need a passport to the Moon), we file into a waiting bus for the trip to the launching pad. The children press their noses to the windows, and even sophisticated travelers are impressed by the sights they are passing. The launch pad superstructure looms ahead and soon we are under it, the giant trusses forming a vast framework over our heads.

We are carried in an elevator up to the passenger door of the shuttle. Because the shuttle is vertical, we are cautioned by the stewardess waiting at the entrance to follow instructions carefully. What appears to be a floor is the front wall of the passenger compartment.

Through the opening in the "floor," we see the seats stacked vertically, How shall we get into them?

A stewardess standing by the opening tells us what to do. Two seats on rails are waiting at the opening. We get in gingerly and the seats slowly move down and stop by our cabin seats. We slide over and strap ourselves into the cabin seats, which tilt forward. Soon, all of the cabin's hundred seats are taken. We peer out of the small elliptical window next to us, a little frightened, exhilarated, and wondering as we were when we took our first ride in an airliner. The Florida sunlight brightens the huge buildings, the giant launching pads, the gantries and runways that even at our height seem to run to the horizon. We look over the blue ocean, the white beach, the green orange groves and the houses far off, the signs of life and people. What are we getting into?

The loudspeaker sounds a warning. Before we realize it, the engines far below us are firing and the gently vibrating ship begins to lift, so slowly at first that we are not aware we are rising until we look down at the rapidly diminishing scene below. The ship accelerates, and it is only a few minutes until we have mounted above the cloud deck and the sky changes from blue to black. The Earth below looks like a map as we watch the Atlantic, Africa, and Asia pass beneath. The television screen on the cabin front wall lights up and we watch our approach to the space station, looking at first like a toy in the distance. As the shuttle slows for docking, we perceive, with a shock, that the Low Earth Orbit Space Station, a confusion of cylinders, beams, modules, spheres, antennas, and docking ports, is over a kilometer long.

Docking is smooth and uneventful. We are told that we will have to wait a few hours in the passenger lounge of the space station. Our guide tells us we can buy souvenirs, films, booklets, photographs, and the like at the souvenir shop in the lounge. We can tour the space station while waiting for the Lunar Spaceship, but we must stay with the tour guide and not wander off by ourselves, for the Lunar Spaceship cannot leave until everyone is accounted for.

We unstrap ourselves and float upward. We are weightless! No wonder the cabin interior is so heavily padded. The stewardesses smilingly tell us to use the handholds to move to the cabin exit. We do so easily, pleased by this novel experience of zero g. Once we clamber out of the shuttle door, helped by the ever-present stewardesses, we are given a pair of magnetized shoes to wear. Clumping around in them in zero g may not be elegant, but the space station staff does not want to contend with hundreds of tourists floating in midair.

Once in the station lounge, the indolent decide to stay in the lounge and watch a wall-sized television screen. We tag along with the energetic, trailing after our station guide who wears her magnetic shoes as if they were slippers. Our first stop is the station exhibit room, where space products are displayed. The silicon strips, vials of vaccines, bars of alloys, metal ingots, and solar cell films are not too interesting except for the captions that explain them. What catches everyone's eye is a huge blue sapphire mounted on a stand in the center of the room, a two-meter (six-foot) shimmering wonder. We move down corridors, up and down escalators and elevators, peering into laboratories, pilot plants, small animal colonies, and rooms filled with incomprehensible black boxes, stopping for frequent short talks by our guide. Some of us would inevitably be left behind but for the two watchful guides who make up the rear guard. We find ourselves back in the lounge and are glad to sit down and relax.

But not for long. The public address system, in that curious flat tone that no advance in technology seems able to cure, announces that the Lunar Spaceship is ready for boarding. Follow the guide, please.

This time the whole group troops after the guide, going down a corridor to the docking port. As we enter the spaceship, our names are called off and are checked off the passenger list as we respond. We find our assigned seats, strap ourselves in, and gratefully slip off our magnetized shoes, stowing them under the seat. The doors slide shut and soon we are gently moving, hardly aware of the faint roar of the engines. We look about. The cabin is much like that of a very large airliner. Since we are lucky enough to have been assigned window seats, we can watch the Space Station diminish in the distance. Then we turn our attention to the black, starry sky. The cabin lighting makes seeing difficult, so we take the light hood out of the pocket of the seat in front, attach it to the windowframe, and slip it on. At first the sky appears pitch black, but as our eyes become dark adapted, stars appear. Soon they shine far brighter than they did from Earth. They do not twinkle but shine steadily, and their colors are vivid, not washed out by the atmosphere. Antares is a glowing red and Sirius a diamond blue. To our delight, we can see all the planets as far as Saturn by looking out of windows on both sides. Some people claim they can see the larger Jupiter moons with their own eyes.

The stewardess's announcement that dinner is being served brings us back to the cabin. We reluctantly take off our light shields and put them away. Dinner on a spaceship turns out to be much like airline fare. After dinner, some of the passengers while away the time with conversation, some read, some play games, and other write letters on tables on their laps. We put on the light hood again, but this time we use our binoculars. Although everyone has seen the sky from space many times on television and in films, nothing can equal the thrill of seeing for the first time the universe unaltered by the atmosphere.

An hour later the cabin lights dim slowly, and most of the passengers slide their seats backward for a few hours of sleep. We follow suit but are soon reawakened as the cabin lights go on. To our surprise, we see by the cabin clock that we have slept for six hours; soon the ship will be ready to land. We begin to feel once more the sensation of weight, although we feel very light. Looking out the cabin window, we see the lunar landscape far below, a surface carved out by giant asteroids four billion years ago.

The public address system warns us to strap ourselves in firmly, and the stewardesses go up and down the aisle to check us. Then they strap themselves into their own seats. The ship is now descending on a steeply slanting course. Its axis slowly turns to the vertical and we are thrust down in our now horizontal seats as the spaceship descends to the landing pad, while the engines make the cabin vibrate as they roar.

We clamber cautiously out of our seats and onto the aisle seats that take us to the cabin exit. It is a relief to stand up straight and feel once more the sensation of partial weight. We walk through the entrance tunnel to the gantry platform and the waiting gantry elevator. A short descent and we are on the lunar surface—more prosaically the passageway floor leading to the reception hall.

A Lunar Base official greets us and tells us to rest awhile to get used to the weak lunar gravity. There are several weighing scales in the hall and everyone weighs in. A man who weighs 890 newtons (200 pounds) on Earth is flattered to find that he weighs only 140 newtons (32 pounds) on the Moon. We are warned by the smiling official not to be fooled, for while weight changes with gravitational fields, mass remains the same.

Everyone is eager to go on the first tour, which will take us around the main lunar base. We are willing to retrieve our luggage later on. We are told that our luggage will be taken directly to our rooms and we need not worry about it. This time, we keep up with our guide readily, having found that walking on the Moon is easy as long as we do not let our steps turn into bounds. Parents have trouble restraining their youngsters. We walk down a long tunnel lit by fluorescent tubes. At its end, the guide stops to let us catch up. He turns *right into a side passage, and we see along the right wall of the passage a glass window that seems to extend as far as we can see. Inside the glass, there appears to be some kind of factory.*

Lunar rocks and soil are being borne on conveyor belts into the mouth of a three-meter (ten-foot)-wide cylinder. They disappear into its depths in a cloud of dust and grinding noise. As we walk along, we see conduits, tubes, panels, and lights interlacing over the cylinder. Our guide explains that we are looking at the central power plant and refinery of the base. The two form a combined plant that is the core of the base's economy. Lunar rocks and regolith are loaded into the receiving end, ground to a fine powder, and passed through magnets to retrieve the iron; the rest of the ore is poured into the interaction section of the fusion torch.[25] We stop and look at it in awe. Inside the cylinder, plasma from the fusion reactor, the central power plant, is vaporizing the rock powder into ions, electrified gases. Separated into their constituent elements by strong magnetic fields, they condense into solids, liquids, and gases, the feedstocks for lunar technology. Farther on, we pause and watch ingots of iron, aluminum, magnesium, silicon, and other elements emerging to be carried away by conveyor belts.

Our next stop takes us to a large dome. In its center stands a gleaming white metal cylinder 30 meters (100 feet) high and 15 meters (50 feet) in diameter. This is the base fusion reactor, which furnishes electricity not only to the base but by underground superconducting cables to outlying facilities.

The silence contrasts with the shattering noises of the refinery. Within these smooth walls, furious fusion reactions at a billion degrees Kelvin are pouring a Niagara of energy into the power system. Energy, the guide declares, is the only true wealth, and he adds proudly that the lunar base has far more energy per capita than any country on Earth, 20 megawatts per person. The lunar colonists are planning to make their colony the most advanced society in the world. They intend it to become the center of the solar system.

We turn back, heading for our starting point, but the guide has one more surprise. At an intersection, he pauses before an open portal. We walk through it and enter an Earthlike hall. The Sun is high in the blue sky, and the green landscape seems to stretch for miles to the horizon. We look up and see that the vast space is covered with a transparent arching roof. The arch is made of fused silica blocks, three meters (ten feet) thick. The blocks interlock and in the lunar vacuum have virtually fused together to form an airtight and immensely strong cover. Their mass and the mass of the air in the hall make an effective shield against the solar wind, solar flares, and cosmic rays. The level of natural radioactivity is lower than on Earth.

The sky is very dark blue, since the arch reaches a height of 500 meters (1600 feet) and there is sufficient air to partially scatter the sunlight. Before us is a park with a farm, playgrounds, flowering gardens, orchards, and recreation centers interspersed. We can see a swimming pool and a diver who bounces off a diving board, floats upward, and slowly drops to the water. The farm surrounds a low rectangular building, which the guide tells us is the CELSS Research Center.

There is so much to see—but the guide declares that it is dinnertime and we can come back later and explore the place at our leisure. Back at the lounge, we are glad to sit once more at tables and eat in appreciable gravity. After dinner the tireless guide relates the history of the lunar base to the accompaniment of a rear-projection screen in the wall we face. The information center, industrial laboratory, and theater are on the agenda for the after-dinner tour, but we are a little tired after the long walk and decide to see them later on. The lunar base runs on the familiar 24-hour Earth day, and our diurnal cycle is set for sleep. The next day we are to take a tour by Lunar Rover to and through the Copernicus Crater, which everyone says is well worth seeing.

Notes

1. *Outlook for Space*, Scientific and Technical Information Office, National Aeronautics and Space Administration (NASA SP-386, Washington, D.C. 20546), p. 7–8.

2. Gerard K. O'Neill, "The Colonization of Space," *Physics Today*, September 1974, p. 36.

3. Myron S. Malkin, "The Space Shuttle—Key Considerations," in *Space Stations–Present and Future: Proceedings of the XXV International Astronautical Federation Congress, Amsterdam, September 30–October 5, 1974* (Elmsford, N.Y.: Pergamon Press, Inc., 1975), pp. 241–259.

4. *Statistical Abstract of the United States*, 100th ed., table, no. 719, 1012. Washington, D.C.: Bureau of the Census, 1979.

5. Peter E. Glaser, "Solar Power from Satellites," *Physics Today*, 30 (February 1977), 30–38.

6. Gordon J. Mac Donald, "An Overview of the Impact of Carbon

Dioxide on Climate," DC-1, American Physical Society Meeting, January 30, 1979.

7. *Newsletter*, Metropolitan Washington Council of Governments, Washington, D.C., January 1978.

8. Marcia C. Smith, "Solar Energy from Space: Satellite Power Stations," Issue Brief No. IB780 12, January 23, 1978, Congressional Research Service, Washington, D.C. 20540.

9. *Nuclear Power*, Edison Electric Institute, EEI Pub. No. 78–24, 1978.

10. J. P. McBride, R. E. Moore, J. P. Witherspoon, and R. E. Blanco, "Radiological Impact of Airborne Effluents of Coal and Nuclear Plants," *Science*, 202:4032 (December 8, 1978), 1045–1050.

11. Bernard L. Cohen, "High-Level Radioactive Waste from Light-Water Reactors," *Reviews of Modern Physics*, 49 (January 1977), 1–20.

12. Bernard L. Cohen, "A Tale of Two Wastes," *Commentary*, 66 (November 1978), 63–65.

13. *The Hydrogen Engine*, Billings Energy Corporation, Fourth Quarter 1976, Box 555, Provo, Utah 84601, p. 7.

14. W. J. D. Escher, "Hydrogen-Fueled Internal Combustion Engine, A Technical Survey of Contemporary U. S. Projects," TEC 75/005, U. S. Energy Research and Development Administration, Washington, D.C. 20545.

15. *Alternative Fuels Utilization Report*, U. S. Energy Research and Development Administration, Washington, D.C. 20545, No. 2 (August 1977), p. 1.

16. "FEF Meeting Reviews Fusion Progress," *Fusion*, November 1978, p. 16.

17. *A Forecast of Space Technology*, Scientific and Technical Information Office, National Aeronautics and Space Administration, Washington, D.C. 20546 (NASA SP-387, January 1976).

18. Dave Dooling, "Super Skylab," *Spaceflight*, 19:8 (July/August 1977), 269–270.

19. *Materials Processing in Space*, National Space Applications Board, National Research Council, Washington, D.C. 20418 (1978).

20. Larry Kramer, "Growing Demand for Communications Satellites Poses Problem," *Washington Post*, January 28, 1979, p. E-1.

21. Ivan Bekey and Harris Meyer, "1980–2000: Raising Our Sights for Advanced Space Systems," *Astronautics and Aeronautics*, July/August 1976, pp. 34–61.

22. Ivan Bekey, "Big Comsats for Big Jobs at Low User Costs," *Astronautics and Aeronautics*, 17:2 (February 1979), 42–56.

23. David Caulkins, "Raw Materials for Space Manufacturing—A Comparison of Terrestrial Practice and Lunar Availability," *Journal of the British Interplanetary Society*, 30 (1977), 315–16.

24. Gregory L. Matloff and Alphonsus J. Fennelly, "A Superconducting Ion Scoop and its Application to Interstellar Flight," *Journal of the British Interplanetary Society*, 27 (1974), 663–673.

25. Bernard J. Eastlund and William C. Gough, *The Fusion Torch*, Division of Research, U. S. Atomic Energy Commission, Washington, D.C. 20545 (May 15, 1969).

7
SOLAR SYSTEM NAVIGATION

To find our way through the solar system, we must navigate—that is, find out where we are and determine where we are going. To do this, we must use a three-dimensional coordinate system. A coordinate system is a way to locate objects in space. For a given purpose, one coordinate system may be superior to others. While two-dimensional diagrams can serve as maps, space navigation is always done in three dimensions.

Coordinate Systems

The rectangular and spherical coordinate systems that we use in our daily lives are also suitable for space navigation. Let us consider the rectangular system first. Many cities and towns in the United States are laid out in grids with streets running north-south or east-west. The origin point of the grid, from which the streets are counted, is at a central location in the city. Let us suppose that we are at the origin point and have an appointment with a friend at 145 East Second Street, N.E., suite 1209. How do we get there? We could walk two blocks east to East Second Street and then north on East Second Street to number 145. Next, we would take the elevator to the twelfth floor and walk to suite 1209.

The address, 145 East Second Street, N.E., suite 1209, specifies a three-dimensional location. One coordinate gives the east-west location: East Second Street; a second gives the north-south location: 145 and N.E.; and the third gives the location relative to street level: suite 1209. (See Fig. 7-1.) As we see, rectangular coordinates are a simple, useful system.

Spherical coordinates are an alternative. Instead of specifying a location by distances along three mutually perpendicular axes, we use two angles that are measured around mutually perpendicular axes and the distance from the origin of the coordinate system. For example, an

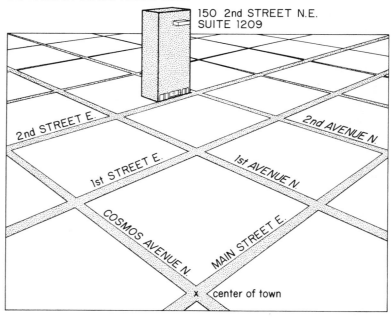

Figure 7-1. Rectangular coordinates for 150 Second Street, N.E., Suite 1209. From the origin in the center of town, one walks two blocks north (first coordinate), then to number 150 (second coordinate), and finally up to the twelfth floor (third coordinate). Suite 1209 refers to the location on the twelfth floor.

airline pilot might specify his position as latitude 40° 15′ North, longitude 75° 20′ West, and altitude 13,100 meters (43,000 feet). He would be using a spherical coordinate system whose origin is the center of the Earth. Latitude is a north-south angle while longitude is an east-west angle. Since the Earth is almost a perfect sphere, we can subtract its radius and use as our radial coordinate the distance above mean sea level. (See Fig. 7-2.)

We could have used the rectangular coordinate system instead of the spherical or vice versa. The choice is a matter of convenience and simplicity. To specify the street address in spherical coordinates, one angle would be the one that the radius vector (the line joining the origin to the address) makes with the East-West line, the second angle would be the angular height of the suite as seen from the origin, and the radial coordinate would be the distance of the suite from the origin. In this example, the spherical coordinate system is much harder to use on a practical basis than the rectangular system. We will meet situations in which the converse is true.

Sometimes a single number will serve for two or more coordinates, or one coordinate ceases to be necessary. To our friend who has reached the twelfth floor of 145 East Second Street, N.E., suite 1209 specifies a location, for the number 1209 combines both east-west and north-south positions.

Figure 7-2. Spherical coordinates. Latitude, longitude, and altitude form three spherical coordinates on the surface of the Earth.

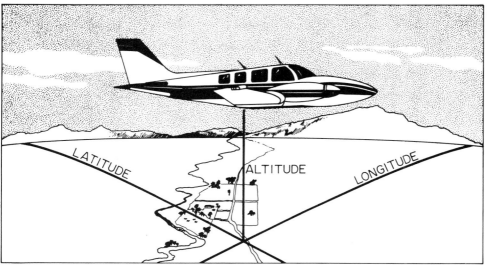

When you fly across the United States, you will notice that the boundaries of farms west of the Appalachians are usually rectangular. They were laid out according to east-west and north-south lines that are arcs of latitude and longitude, which are spherical coordinates.

Planetary Coordinates

The space navigator planning a course to a planet such as Mars will specify the landing site. What coordinate systems, then, are suitable for locating positions of spherical bodies such as planets and moons? Let us look at one of these bodies as a space traveler would see it. All of them rotate about an axis. From our space traveler's viewpoint, surface features move smoothly across the visible hemisphere as it rotates. By photographing or making accurate drawings of the hemisphere and recording the times of our observations, we can determine how long it takes surface features to return to some arbitrarily chosen positions on the visible hemisphere and the apparent paths these features make. The time interval it takes a given feature to return to its initial position is the period of rotation, while our examination of the paths reveals the location of the poles of rotation. By observing many rotations, we can determine the period and the location of the poles more precisely. Also, compensation has to be made for any change in our vantage point. We have assumed that the surface features do not change their positions relative to each other. If they do, suitable corrections can be made.

We can use the poles of rotation to set up a coordinate system. (See Fig. 7-3.) Any circle on the surface of a sphere whose center coincides with the center of the sphere is called a *great circle*. The great circle called the equator is halfway between the poles. Let us adopt the convention that the north pole is the one about which the hemisphere turns counterclockwise if the observer is above it. The opposite pole is the south pole. From an external vantage point, if north is up and south is down, then east is right and west is left.

Meridians are great circles that pass through the poles and intersect the equator at right angles. Meridians are used to specify east-west locations.

From what point are east-west positions measured? Any distinguishable feature's meridian will do. On Earth, we use as the zero-degree meridian the one that passes through Greenwich, England,

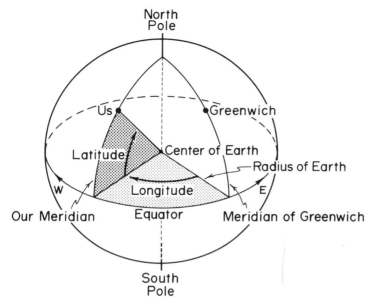

Figure 7-3. Earth-bound geographic coordinates. The origin of this system is the Earth's axis of rotation. Its axis intersects with the Earth's surface at the North and South Poles. The Equator is halfway between them on the Earth's surface. The poles are joined by meridians that can pass through any place on the surface of the Earth. The latitude of a place is the angle from the place to the Equator, as seen from the center of the Earth. The longitude of a place is the angle between its meridian and Greenwich meridian, as seen from the center of the Earth.

where the old Royal Observatory was located. For Mercury, east-west positions are measured from the 20-degree meridian, which passes through the center of the crater Hun Kal (Mayan for "twenty").

The *longitude* of a place is the angle (in degrees, minutes, and seconds of arc) between the meridian passing through the place and the reference meridian as seen from the center of the object. Longitudes are usually measured to the west of the reference meridian, except on the Earth, where for historical reasons they are measured both to the east and the west of the Greenwich meridian.

The *latitude* is the second angle needed to specify the location of a place on a spherical surface. It is the angle (in degrees, minutes, and seconds) to the place measured from the equator along the local meridian as seen from the center of the object. Latitudes are measured both north and south of the equator from 0° to 90°. North is positive and south is negative.

The third coordinate needed to locate a place on a spherical or nearly spherical surface is its distance from the center. It is usually measured as the height or depth with respect to some mean level.

More complicated coordinate systems have to be used for markedly nonspherical bodies such as the moons of Mars, Phobos and Deimos, which are roughly ellipsoidal in shape, something like a football.

The fundamental problem in displaying the surface area of a sphere is that it must be distorted when transferred to a flat surface. A small part of a spherical surface can be shown as a flat map with reasonable accuracy, but large parts require some type of projection.

Circles connecting places with the same latitude are called *parallels*. They are perpendicular to the meridians. Equally spaced meridians and parallels are often used in maps. On a sphere, however, the meridians are of equal length while the parallels decrease in size toward the poles. In any projection, only a selected set of meridians or parallels can be true—that is, have the same length that they would on a globe of the same size. All the other lines will be too long or too short as a result of projection. A good example is a map of North America on a Mercator projection. Canada looks dramatically larger than the United States, although the two countries are about the same size.

Equatorial Coordinates

Now we will extend the technique for specifying positions on planets to locating objects in space, which is basic to space navigation. Looking up at the sky, we see a great hemisphere stretching above us. With a little imagination, we can assume that the sky is an enormous sphere, only half of it above the horizon at any time. The ancient Greeks believed that the stars were attached to such a *celestial sphere*, which they imagined was beyond the orbit of Saturn. The celestial sphere rotated on its axis once a day and thereby moved the Sun, the planets, and the stars across the sky.

While we know that the Earth's rotation causes celestial objects to change their apparent positions, astronomers retain the idea of a celestial sphere, but now they say it has an infinite radius. This amounts to projecting all of space onto a spherical surface, or specifying the direction of a celestial object while ignoring its distance. Also, the infinite radius means that an observer anywhere in the universe can consider himself to be at its center.

Equatorial Coordinates 111

We encounter this projection every day. When we photograph a scene, we are projecting three dimensions onto a flat two-dimensional surface. With a small fraction of a hemisphere, there is little distortion, but if we want to record a large part of a hemisphere without distortion, we must do so onto a curved surface. The curved screens and multiprojectors of wide-angle movies are a good illustration.

Various astronomical coordinate systems are used to specify positions on the celestial sphere. In the equatorial system, right ascension and declination are the analogs of longitude and latitude. (See Fig. 7-4.) This system, designed for Earth-based observing, is the prototype for related systems that will be used for astronomical observing from other celestial objects and from space. Coordinates can be transformed from one of these systems to another by the use of algebraic expressions.

Figure 7-4. The equatorial system. This astronomical system is analogous to the Earth-based geographic coordinate system (Figure 7-3). The celestial poles are analogous to the geographic poles, the hour circles to the meridians, the declination to latitude, and the right ascension to longitude. The vernal equinox is analogous to the intersection of the Greenwich meridian and the Equator. The equatorial system is useful for measuring positions on the celestial sphere as seen from the Earth.

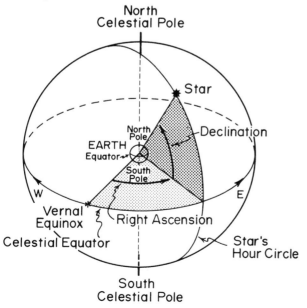

The celestial sphere appears to rotate about points located directly above the Earth's north and south poles. These points are called the *celestial poles*. The *celestial equator* is a great circle on the celestial sphere halfway between—or 90 degrees away from—the celestial poles. The celestial and terrestrial equators lie in the same plane. The analogs of the meridians are the *hour circles*, which pass through both celestial poles and intersect the celestial equator at right angles. The analog of latitude is *declination*; the declination of a point on the celestial sphere is the angle between the point and the celestial equator along the point's hour circle as seen by the observer. *Right ascension* is the analog of longitude, and its reference point is the vernal equinox. This is one of the two points on the celestial sphere where the celestial equator and the plane of the Earth's orbit intersect. Right ascension is measured eastward from the vernal equinox along the celestial equator to the point's hour circle.

The Earth's rotation axis is very slowly changing its direction in space over a period of 26,000 years. This motion is called *precession*. As a consequence, the equinoxes move very slowly around the celestial equator. To specify the location of an object on the celestial sphere, we must give the right ascension, declination, and time (epoch) of the observation. As these coordinates change very slowly and predictably, values can be transformed from one epoch to another.

The equatorial system makes it easy to track any object across the sky. Most astronomical telescopes are mounted with mutually perpendicular axes, one for motion in right ascension and one for motion in declination. One axis points to the celestial poles and is therefore parallel to the Earth's rotation axis. The telescope can be set to point at any object's right ascension and declination, with the motion of each coordinate independent of that of the other. A small motor drives the polar axis (the axis pointing to the poles) at the rate at which the Earth rotates with respect to the stars and so keeps the telescope pointing at the object's right ascension and declination. The equatorial system is suited for Earth-based astronomical observing. For observing on other planets, we will use similar systems by defining the coordinates in terms of the planet's axis of rotation.

Ecliptic Coordinates

There is no preferred axis of rotation when making observations for space navigation out in space. The equatorial system can be used, or adapted, or a new system set up. A similar system, the Sun-centered ecliptic system, is better for determining the positions of planets and therefore for solar system navigation.

Since the orbital planes of the major planets of the solar system are only slightly tipped with respect to one another, the plane of the Earth's orbit about the Sun is only slightly tilted from the mean plane of the solar system. The planets move close to the *ecliptic*, the apparent path of the Sun, which is the reflection of the Earth's motion around the Sun onto the celestial sphere. The ecliptic system is useful for locating the positions of the planets, since most of their motion is in celestial longitude.

In the ecliptic system, the ecliptic corresponds to the celestial equator of the equatorial system. The ecliptic poles are perpendicular to the plane of the Earth's orbit. Since the Earth revolves around the Sun in the same direction that it rotates on its axis, the north ecliptic pole is in the same hemisphere as the north celestial pole, although (because Earth's axis is tilted) the two north poles are 23½ degrees apart. Celestial latitudes are measured from the ecliptic (zero degrees latitude) along great circles that pass through the ecliptic poles. Celestial latitudes toward the north ecliptic pole (90°) are positive, while those toward the south ecliptic pole ($-90°$) are negative. Celestial longitudes are measured from the vernal equinox to the east along the ecliptic, just as right ascension is to the east along the celestial equator.

For solar system navigation, a coordinate system whose origin is the Sun is clearly suitable. This is the Sun-centered modification of the ecliptic system. The mean plane is that of the solar system instead of the ecliptic, and the reference point for longitude is also the vernal equinox. (See Fig. 7-5, page 114.) This system has a major advantage for solar system navigation, for it simplifies calculating the positions of all bodies in orbit around the Sun.

Finding Distances

Besides finding the direction of an object, the space navigator must also know its distance. The basic method for finding distances to distant objects is triangulation, which is used by surveyors and navigators on Earth. Consider triangle *ABC* in Fig. 7-6 on page 114. To determine a triangle, you need only know the lengths of the two sides and the included angle or two angles and the included side. Suppose that C represents an Earth satellite. The distance to C can be determined by

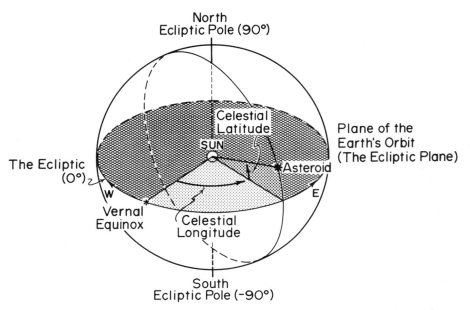

Figure 7-5. Sun-centered ecliptic system. This astronomical system is analogous to the Earth-based geographic system (Figure 7-3.) Its center of origin is the center of the Sun, its plane is the ecliptic, the plane of the Earth's orbit. Its poles are the poles of the ecliptic.

Figure 7-6. Triangulation. Observers at A and B see the same object, such as a satellite, at C, from different directions. The difference in angle, or angular displacement, is ∠ ACB.

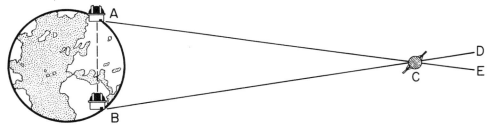

observing it from A and B, determining the distance AB, and measuring the angles ABC and BAC. BC and AC are found by trigonometry.

This method is the basis of depth perception. Our base line is the distance between our eyes. We judge distances to objects we see by angular displacement—that is, the angle ACB, which is called the *parallax*. The closer the object, the larger the parallax. We learn this method of distance judgment unconsciously as infants.

Another way to determine distance is based upon the speed of light. A laser is attached to a large telescope, which is aimed at a retroreflector such as one left on the Moon by the Apollo astronauts.

A collimated high-intensity light beam is transmitted in short pulses, and the reflection is detected. The time between sending the signal and receiving the return echo corresponds to twice the distance covered. The principle is the same as that of radar. To date, laser ranging has been used only with the Moon, while radar has been used for the planets of the inner solar system.

If a moving spacecraft transmits at a constant frequency to, say, an Earth-based station, then the received frequency will not be the same as the transmitted frequency, because of the relative motion of the spacecraft and the receiver on Earth. The frequency difference is related to the relative velocity along the line of sight. By keeping track of the frequency shift as a function of time, we can determine the separation between the spacecraft and the receiver very precisely.

Spacecraft Trajectories

Distance determination and position location are the basis of navigation, the art of arriving at a desired position at a desired time, as planned. If we are traveling from the Earth to Mars, for example, we will not aim our spaceship at the point where Mars appears to be in the sky; rather we will aim at the point where Mars will be when the ship arrives in its vicinity. To plan our trip, using the methods of celestial mechanics, we will need to know not only the movement of Mars but also the positions of the major planets, for their gravitational fields affect the orbit of Mars. Also, we will need to determine the true trajectory of our spaceship and be able to make necessary corrections en route so it can reach its target.

In recent years, radar and unmanned interplanetary spacecraft have greatly improved the accuracy of our knowledge of the distances of the planets. Up to 1960 or so, the accuracy of predicted planetary positions for the inner planets was about 500 kilometers (300 miles). The uncertainty has been reduced to about 50 meters (150 feet). When a spacecraft is sent on a flyby past a planet, the Earth-based determination of its mass and the prediction of its future postions can be greatly improved. This makes planning future spacecraft missions to the planet much easier. Measurements of the positions of the planet's satellites by television cameras on passing spacecraft greatly improve the Earth-based orbit predictions. This is particularly important in investigating the satellites as well as in planning the trajectory through the satellite system. By 1982, refinements in interplanetary navigation and spacecraft flights past Jupiter and Saturn should advance navigation to them to the level of accuracy attained for Mars.

The forces acting on a spacecraft, other than those due to its own propulsion, can be classified as gravitational and nongravitational. The former are far stronger, but the latter are harder to include in calculating predicted trajectories. One that is tractable is the pressure of solar radiation on the spacecraft, although the precise value fluctuates with changes in the solar wind. Besides that, the reflectivity of the spacecraft changes, and such changes are hard to predict. Also, the spacecraft fires gas jets to maintain its desired orientation in space. These jets cause small accelerations which result in velocity changes. A typical interplanetary probe is moved by these nongravitational forces a few meters a day from its planned path. As spacecraft grow in mass, these forces may become less important.

The gravitational force on the spacecraft is the resultant of all those forces between the spacecraft and the bodies of the solar system. Gravity is an attractive force whose strength is proportional to the product of the masses of the two interacting bodies and inversely proportional to the square of the distance between them. If we halve the distance between two objects, the gravitational force between them quadruples, while if we increase their distance three times, the force decreases to a ninth of what it was. Isaac Newton, the discoverer of the law of universal gravitation, found that spherical bodies act as if their mass were concentrated at their centers. Therefore, the distance used in computing gravitational forces between these bodies is taken from center to center.

The Sun is by far the most massive body in the solar system, so when a spacecraft is not near a planet, most of the gravitational force exerted on it comes from the Sun. The planets exert much less force, with Jupiter the main contributor. The planets, moons, asteroids, and comets are gravitationally dominant in their vicinities, the extent depending upon their mass and upon their distance from the Sun and from other bodies in the solar system.

Early in the seventeenth century Johannes Kepler discovered three basic laws of motion in orbit, which Newton later showed were consequences of the law of universal gravity and his three laws of motion. Kepler's first law says that each planet has an orbit around the Sun that is an ellipse with the Sun at one focus. An ellipse is a curve that has two fixed points, the foci. The sum of the distances from the foci to any point on the ellipse is constant. The circle is a special case of the ellipse in which the foci coincide. The semimajor and semiminor axes of an ellipse are one-half its longest and smallest dimensions. Except for Pluto and Mercury, the planets have elliptical orbits that are almost circular.

Kepler's second law says that a straight line joining the Sun and a planet sweeps out equal areas in equal intervals of time. The planet moves fastest when it is closest to the Sun and most slowly when it is farthest away. It speeds up as it approaches the Sun and slows down as it recedes from it. Kepler's third law says that the squares of the orbital periods of the planets are proportional to the cubes of the semimajor axes of their orbits. Therefore, the closer the planet is to the Sun, the shorter is its period.

Strictly speaking, Kepler's laws apply just to two-body interactions. Nevertheless, they serve as good approximate guides to the motions of planets about the Sun, of satellites about planets, and of spacecraft in the solar system. If an object has enough energy to escape from the solar system, its orbit will be a hyperbola, which is an open curve, instead of an ellipse. When calculating a precise trajectory, we compute the forces acting on the spacecraft at a number of points on the trajectory. The problem is actually a multibody one, as the spacecraft interacts gravitationally with all the bodies in the solar system. Once a spacecraft has accelerated away from its launch planet and is en route to its target, it follows the sector of an elliptical orbit that it would complete if the target were not there at the predicted time.

A spacecraft can follow any of a wide range of orbits from launch point to target. The choice depends upon the amount of energy the spacecraft's engines can expend and the object of the mission. The *least-energy orbit* is the one that requires the least expenditure of energy to follow. In the early days of space flight, least-energy orbits were favored, since rocket engines were comparatively weak.

The least-energy orbit from the Earth to Mars is an ellipse tangent to the Earth's orbit at the spacecraft's perihelion (closest point to the Sun) and tangent to the Martian orbit at aphelion (farthest point from the Sun). The spacecraft is launched from the Earth so that at its time of arrival, about 8½ months later, Mars will be on the side of the Sun opposite the Earth's position at the launch date. (See Fig. 7-7.) The time required is one-half the time the spacecraft would take to complete its orbit. The least-energy orbit from Earth to Venus is similar. The aphelion point is now tangent to the Earth's orbit, and the perihelion point is tangent to the orbit of Venus. The spacecraft is launched about five months before Venus will be on the other side of the Sun from the Earth's position at launch date. (See Fig. 7-8.)

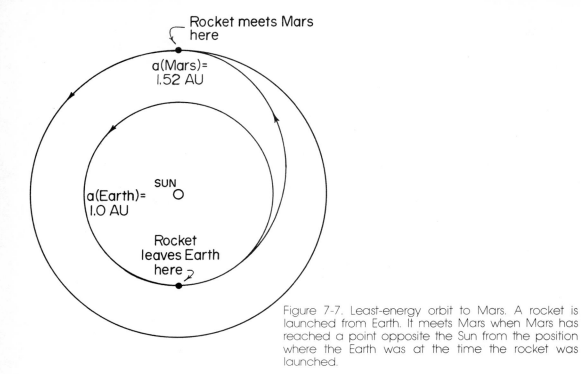

Figure 7-7. Least-energy orbit to Mars. A rocket is launched from Earth. It meets Mars when Mars has reached a point opposite the Sun from the position where the Earth was at the time the rocket was launched.

Figure 7-8. Least-energy orbit to Venus. A rocket is launched from Earth. It meets Venus when Venus has reached a position opposite the Sun from the position where the Earth was when the rocket was launched.

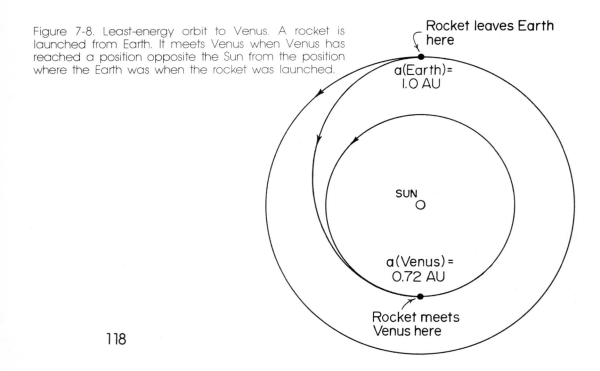

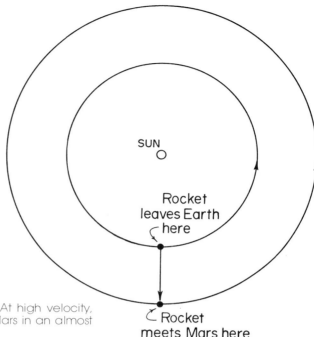

Figure 7-9. High-energy orbit to Mars. At high velocity, a rocket can travel from the Earth to Mars in an almost straight line.

Least-energy orbits have two major disadvantages: they are the longest possible usable orbits and they are very restrictive on possible launch dates. They are, in effect, "poor-man's" orbits. With better engines we can cut down the travel time as well as widen the launch period. The faster the spacecraft goes, the straighter its trajectory can be. (See Fig. 7-9.) The more energetic orbits for the outer planets are sections of ellipses whose perihelion points are closer to the Sun than the Earth and whose aphelion points are farther away than the target planet. For the inner planets, the perihelion points are closer to the Sun than the target planet and the aphelion points are father away from the Sun than the Earth's orbit.

If a spacecraft uses its engines to accelerate during its flight, then it changes from one Keplerian orbit to another. When it reaches the vicinity of its target planet or moon, its trajectory will be altered by the gravitational interaction between the spacecraft and the target. If the spacecraft is to go into orbit about its target, it must decrease its velocity to below that of the target's escape velocity, which it does by firing its retrorockets. The spacecraft then becomes a satellite of the target planet or moon. The spacecraft's new orbit can be changed in shape or have its altitude increased or decreased by additional retrorocket burns. Deviations from the predicted orbit furnish important information that is used to improve the orbital calculations. They can also show the difference of the planet or moon's shape from a perfect sphere and any nonuniform distribution of its mass.

Reference on present interplanetary navigation capabilities and suggested for further reading on this subject is William G. Melbourne, "Navigation Between the Planets," *Scientific American*, 234:6 (June 1976), 58.

8
SOLAR SYSTEM TRAVEL

Space technology has passed the pioneering stage. It has reached the level of maturity at which public preferences for certain types of missions can and should begin to be considered. For example, a lunar geologist, in planning his trip to the Moon, would like to spend as much time as possible on the Moon itself rather than in traveling back and forth. If space tourism is to become popular, the wishes of the public must be similarly considered. Let us view solar system travel from the viewpoint of the average American.

The differences between the Space Shuttle and the solar system spaceship will be due to their different mission requirements. The Space Shuttle is designed for close-in missions to low Earth orbit. It cannot even travel to the Moon, since its engines cannot achieve Earth escape velocity. The Space Shuttle is a great advance, but it is only the first step toward the colonization of the solar system.

The first railroad train revolutionized land transportation by shifting it from animal power to mechanical power. Railroad cars were improvements on stage coaches, and the public perceived them as such. When railroads were introduced in the 1830s, people eagerly gave up stage coaches for railroad trains. For ocean travel they gave up sailing ships for steamships. When the airlines began large-scale transatlantic service after World War II, the ocean liners were doomed. Why did these changes take place?

The railroad train and steamship offered both greater comfort and higher speed than the stage coach and the sailing ship. Airliners, however, even the largest, offer little more than legroom and a stroll up and down the aisle. Speed is the sole reason for their success. For reasons of speed, and despite the vehement opposition of the environmentalists, the Anglo-French supersonic airliner has firmly established itself on the transatlantic route.

The lesson of the last century and a half is flatly that most people regard vehicles as a means of getting from one place to another—and the faster, the better. People want to spend their available free time at their destination, not en route. This will hold true for space travel. When spaceships fly to the Moon, the spaceship that takes three days to get there will quickly be deserted for the new one that takes only one day.

How do we want to travel in the solar system? Do we want interplanetary voyages to compare in duration with trips on Earth? If so, consider what will be required of space technology.

The Moon is only 384,000 kilometers (a quarter of a million miles) from the Earth. The Apollo spaceships reached it in four days. That may seem fast, but not when measured against the scale of the solar system. Pluto, the most distant planet from the Earth, is on the average 15,000 times as far as the Moon. At Apollo speed, it would take 191 years to travel to Pluto. But for most people even a two-year trip to Pluto would be too long. Solar system spaceships, therefore, will have to travel at an average speed at least a hundred times that of the Apollos.

Apollo 11 was launched on July 16, 1969, at 9:32 A.M. EDT. The Lunar Module landed on the Moon on July 20 at 4:18 P.M. after a flight of 103 hours, having averaged about 4000 kilometers (2500 miles) an hour. If the S-IVB's engine had imparted to the spaceship a parting velocity of 153 meters (500 feet) a second more than it did, the Apollo would have reached the Moon in two days instead of four.

As we have seen in Chapter 6, a trip to the Moon in one day is feasible with chemical rocket engines. The average speed would be 16,000 kilometers (10,000 miles) an hour, four times that of Apollo 11.

That is fine for a lunar trip, but would it do for solar system travel? Not at all. At this speed a spaceship would take 387 days to cover an astronomical unit of 149,500,000 kilometers (92,960,000 miles). A voyage to Pluto at its aphelion would require 53 years and 3 months, while a trip to Mars would take 2½ months. This calculation does not take into account the gravitational fields of the planets and the curving trajectory of the spaceship, which would lengthen its trip.

We will make a reasonable prediction. People will want to spend only a few weeks at most in traveling to the planets. No matter how fascinating the sights, most of them will not want to spend months or even years en route to the satellites of Jupiter or Saturn. Speed will be a chief requirement for mass solar system travel.

There is an exception to this prediction. Space explorers will be willing to put up with far longer voyages than the space tourist or the traveler going on an assignment, such as construction of a space base. Many of the expeditions sent out in the sixteenth to eighteenth centuries lasted as long as three years or even more, and the sailors had to endure living in conditions that we regard as intolerable. Space explorers probably will live under conditions far less comfortable than those of the spaceliners that will follow, once the planets and moons have been explored and the bases established, but they will be willing to do so. Their compensation will be the excitement of discovery, the sheer adventure of exploring new worlds that no human eyes have ever seen before. Imagine how it will feel to land on Mars for the first time in human history or to orbit Io and watch its enormous volcanoes spewing glowing clouds into black space.

Mercury, Venus, and Mars are less than one astronomical unit from the Earth when they are on the same side of the Sun as the Earth, while the outer planets are much farther away. At its closest, Jupiter is over four astronomical units distant, Saturn over 8.5 astronomical units, and so on. The problem is how much time a solar system spaceship should take to travel one astronomical unit. To make solar system travel popular, a spaceship would have to do about 320 kilometers (200 miles) a second. At this velocity it will traverse an astronomical unit in five and a half days. Table 8-1 shows the travel times to the planets from Earth at this speed. Also shown is the average distance of each planet from the Sun (the semimajor axis of its orbit) in astronomical units. The trip times for Mercury and Venus are for inferior conjunction, when they are between the Sun and the Earth—that is, approximately their closest approach. The trip times for Mars and the other outer planets are for opposition, when the Earth is

TABLE 8-1. Interplanetary Travel

	Planet's Distance from the Sun (AU)	Trip Time
MERCURY	0.387	3 days, 7 hours
VENUS	0.723	1 day, 12 hours
EARTH	1.000	—
MARS	1.524	2 days, 20 hours
JUPITER	5.203	22 days, 17 hours
SATURN	9.539	45 days, 23 hours
URANUS	19.182	97 days, 20 hours
NEPTUNE	30.058	156 days, 8 hours
PLUTO	39.518	207 days, 5 hours

on a line between the planet and the Sun—again about their closest approach.

These figures are more realistic than they might seem at first glance. Acceleration and deceleration would not take much time. Accelerating at one *g*, a solar system spaceship could reach 320 kilometers (200 miles) a second in only 9 hours and 7 minutes, so the time spent in gaining cruise velocity and in slowing down to landing would be relatively minor. Also, at this speed the gravitational fields of the planets would have little effect upon the spaceship's course. Instead of following a curving trajectory, it would fly in almost a straight line from planet to planet.

Let us make the reasonable assumption that the spaceship's engines have an exhaust velocity about equal to the spaceship's speed. Its engines then must have a specific impulse of about 32,800 seconds. This immediately eliminates chemical engines, which have a maximum specific impulse of about 500 seconds, and the solid-core nuclear engine with its probable maximum of about 1000 seconds. Only the gas-core and fusion engines are left as possibilities. NASA studies contemplated a first generation of gas-core engines with a specific impulse of about 2000 seconds, which would leave quite a way to go. If true solar system spaceships are to become reality, then the engineers must develop and test a series of successively better gas-core engines, just as they have with the hydrogen-oxygen rocket engine, resuming work on the prototype and reaching the desired specific impulse by, possibly, the third or fourth generation. The effort will probably cost billions—and will be well worth it.

It is evident that there will be a marked difference between flights to the Moon and those to bodies out in the solar system. Lunar travel will come to resemble air travel. The shuttles and the lunar spaceship, since their trips can be counted in hours, will have passenger cabins filled with rows of seats, and their accommodations will resemble those of long-distance airliners. In time, people may even commute biweekly to the lunar colony. This is not at all impossible. Sixteen kilometers (10 miles) per second equals 58,000 kilometers (36,000 miles) an hour. If we allow for acceleration and deceleration, a trip to the Moon would take about nine hours. Some American businessmen are now working in Japan while living in the United States. Their biweekly commuter flights take them more than nine hours.

Travel to the planets will be far more arduous than travel to the Moon because of the far greater distances. A three-day trip to Mars at 320 kilometers (200 miles) a second will not be easy to achieve. For a long time before that stage is achieved, the Mars voyage will take months. The Mars colony will certainly have to be more self-sustaining and more independent of Earth support than the lunar colony.

A solar system spaceship will resemble an ocean liner more than an airliner. Of necessity, it will be much larger than the lunar spaceship. Assembled at a space station, it will remain in space throughout its life history. The engines will be fed by huge tanks capable of holding the vast volumes of hydrogen needed for acceleration and deceleration. Also, once solar system spaceships are in operation, there will be every reason to make them as large as feasible. Founding and supporting planet and moon colonies will require large payloads and many colonists. If sufficient power can be provided, the cost per kilogram of payload goes down with increasing size. The small size of our unmanned interplanetary spacecraft today is a reflection of the immaturity of space propulsion. Lunar spaceships may well be 61 to 91 meters (200 to 300 feet) long, while solar system spaceships may reach 300 meters (1000 feet) in length.

The public's tastes will influence the design of solar system spaceships as they have ocean liners and airliners. Besides being fast enough to voyage to the planets in acceptable time, they will have to be large enough to be reasonably comfortable as well as reliable. The latter desire will be readily satisfied. With the aid of sophisticated astrodynamics, celestial mechanics, and advanced navigational instruments, the space navigator will guide the course of his spaceship with remarkable precision. He will also have the help of a network of radars and radio beacons scattered through the solar system. Besides, except for solar flares, there are no storms in space, no real dangers of collision, no reefs or shallows. Meteoroids have turned out to be much less a hazard than had been feared, and radar should prevent any accidental encounter with a large meteoroid or small asteroid. Space-lines may be able to schedule their spaceships' departures and arrivals to the minute, even for distances of hundreds of millions of kilometers.

Space travel by spaceliner may be the most luxurious form of travel in history. Sea-level air pressure will be maintained, so there will be no popping of ears, which can be painful to many travelers. Acclerations may be less than the three g's of the Space Shuttle. Temperature will be equable and could be varied smoothly from "day" to "night" to avoid monotony. The continual noise and vibration of even the latest airliners will be absent. Vibration and noise will probably cease once the engines have stopped, and they will be off at least 90 percent of the time for interplanetary voyages. The stillness will be relieved only by the faint hum of fans, motors, and other equipment, except when engines are fired for midcourse corrections. The spaceliner will regain spaciousness for the traveler who has traveled in cramped airliners. Besides passenger cabins and dining rooms, it will have lounges, recreation rooms, and observatories for viewing the heavens.

When ocean liners were being taken off the transatlantic routes, some were kept in service by being sent on cruises to the Caribbean, the Mediterranean, and other vacation areas, and this led to a new kind of vacation cruise. Someone had the bright idea that enough people might be interested in a cruise in pursuit of a solar eclipse to make it worthwhile. To the astonishment of the travel agents, the cruise turned out to be highly popular. Soon there were dozens of science cruises every year for people interested in astronomy, marine biology, oceanography, ecology, and much else, who absorbed lectures by scientists as an earlier generation of vacationers had pursued bridge while crossing the Atlantic. So, too, there will be science space cruises, tours of the Moon, Mars, Jupiter (at a safe distance), and other planets and satellites, replete with lectures, observation, discussions, workshops, and much amateur picture taking.

Will space travel have the powerful appeal of Earth travel? We can make an estimate by comparing their offerings. Why do millions of Americans spend their vacations each year touring Japan, France, Switzerland—indeed, every country on Earth? The explanation is that we all like to experience the actuality of travel—to see for ourselves. Seeing foreign lands on television or in films or by reading books only whets one's appetite for going there. More precisely, we go overseas to see at first hand the works of Nature and the works of Man.

Our own globe has features not found anywhere else in the solar system, notably bodies of liquid water and two million species of living beings. The planets, moons, asteroids, and comets, however, will present sights unparalleled on Earth. The space traveler will look forward to touring space stations, lunar bases, and planetary colonies. Over the decades, the colonies on far-off planets may well evolve along individual lines, even though advancing space communications and transportation will keep them from evolving into vastly different cultures. By no means, at any rate, will the blasé space traveler be able to say that if you've seen one planet colony, you've seen them all.

Exploring will be the prime motif of space touring. The space enthusiast will be able to climb mountains (Olympus Mons on Mars is 22,800 meters [75,000 feet] high), ride in lunar rovers over Mare Imbrium, follow the vast escarpments of Mercury, hundreds of kilometers in length, and even cross glaciers such as the Martian North Polar Cap. We have just begun to discover the wonders of the solar system. The future space traveler will have an area of terrain several times that of the Earth awaiting his or her personal exploration.

9
EXPLORING AND COLONIZING MARS

What is Mars like? When Galileo, the first scientist to use a telescope, looked at Mars through his primitive instrument in 1610, he saw a blurred red disk with no detail. Telescopes were slowly improved but it was not until 1659 that Christian Huygens, a Dutch physicist, published a map of Mars showing a feature, Syrtis Major, that we recognize today. Not until the 1800s, however, did telescopes become capable of resolving the major features of the Martian disk.

Giovanni Schiaparelli, an Italian astronomer, observed Mars carefully at five oppositions from 1877 to 1886. At an opposition, an outer planet is on a direct line from Sun to Earth to planet and so is as close as it gets to the Earth. The rules that Schiaparelli devised for naming the features of Mars are still largely followed. In 1877 Asaph Hall of the U. S. Naval Observatory discovered the two tiny moons of Mars, which he named Phobos and Deimos. In the early 1900s Percival Lowell, an amateur American astronomer who became a professional, established Lowell Observatory in Arizona to observe Mars. He became convinced that Mars was covered by straight canals connecting oases and that the canals had been dug by intelligent beings. His theory was later disproven by Antoniadi and other astronomers.

By 1964 considerable progress had been made in observing Mars, but knowledge of the planet was still sketchy. The diameter of Mars had been determined to be 6720 kilometers (4170 miles). Its axis is inclined 25.1 degrees from the perpendicular to its orbit, a little more than the Earth's 23.5 degrees. Like the Earth, Mars has seasons, but they last about twice as long since the Martian year is 687 Earth days long. The north pole of Mars points to Deneb, not to Polaris as the Earth's does. Martian gravity is about 38 percent of the Earth's gravity, so a man weighing 900 newtons (200 pounds) on Earth would weigh 340 newtons (76 pounds) on Mars. The Martian day is called the sol. It is 24 hours, 37 minutes, and 22.4 seconds, not much longer than our Earth day.

Since, on the average, Mars is 1.52 times as far from the Sun as the Earth, it certainly should be colder. The solar constant, the heat received from the Sun each minute by a square centimeter at the top of the atmosphere, is about 2.00 calories for the Earth, while that of Mars ranges from 0.817 to 0.833 calories. Mars therefore receives about 43 percent as much sunlight as the Earth, because of its greater distance from the Sun. In 1927, Coblentz and Lampland of the National Bureau of Standards measured the infrared radiation from Mars and found that the planet had an average surface temperature of 245°K (−18°K).

Spacecraft Discoveries

The abrupt transition from telescopic Mars to spacecraft Mars came on July 14, 1965, at 5:30 P.M. Pacific Standard Time via the Goldstone radio antenna of the Jet Propulsion Laboratory, when the message arrived that the Mariner IV spacecraft had begun to take pictures of Mars 12 minutes earlier. At that time, Mars was 12 light-minutes from Earth. As Mariner IV swung around the planet, it took and later sent 21 photographs. The closest was taken at a distance of 11,850 kilometers (7400 miles). The edge of the globe was outlined by the dark sky, and vague grey and white areas could be seen on its surface. The Mariner team was astounded to see craters everywhere.

Although the craters came as a surprise, they had been predicted by a prescient astronomer. In a paper published in 1950, Dr. Ernst J. Öpik of the Armagh Observatory wrote[1]:

> With the scarcity of water and the lesser density of the Martian atmosphere, erosion should proceed there much slower than on earth; the trace of impact should stay at least for 10 million years and, perhaps, ten times longer than that, in which case the surface of Mars should be covered with hundreds of thousands of meteor craters exceeding in size the Arizona crater.

Although the pictures showed less than 1 percent of the Martian surface, more than 70 craters were found in them. If Mars is, indeed, so crater packed, it might have more than 10,000 craters from 5 to 120 kilometers (3 to 75 miles) across. As Mariner IV passed Mars, its radio beam penetrated the Martian atmosphere on its way to the Earth. Scientists determined that its atmospheric pressure was only 4 to 10 millibars, about the atmospheric pressure on Earth at an altitude of 30,500 to 36,500 meters (100,000 to 120,000 feet). (The Earth's sea-level pressure is 1013 millibars.)

The planetologists compared the Mars pictures with those of the Moon and concluded that the Martian terrain was 2.2 to 3 billion years old, since the distribution in the range of size of the Martian and the post-mare craters on the Moon was similar. Apparently, about half of the Martian craters have been destroyed by erosion since their formation. If the rate of impact of asteroids on the Moon and Mars has been constant since the formation of the solar system about 4.5 billion years ago, the topography of Mars appears to be about half as old as that of the Moon.

Both the scientists and the public were discouraged by the Mariner IV discoveries. Mars, it seemed, was almost as hostile to life as the Moon. Nothing on Earth can live at an altitude of 30,500 meters (100,000 feet). Belief in life on Mars seemed to be as fanciful as the novels of Edgar Rice Burroughs. Mars was disappointingly lunar—no mountain ranges, no canyons, no rivers, no sea basins. No canals were seen on the Mariner IV photographs. The dream held for three centuries that Mars was a companion of Earth had been shattered.

Even so, what had been learned was priceless. NASA continued its exploration of Mars. In the fall of 1965, work began on Mariner Mars 1969, which was to answer the questions raised by Mariner IV.

In 1969 Mariner VI and Mariner VII took hundreds of pictures far sharper and clearer than those of Mariner IV. Hundreds of craters were evident but no mountain ranges or valleys. Their instruments reported that the temperature was up to 15°C (60°F) near the equator but about −150°C (−240°F) over the frozen wastes of the south pole. Air pressure ranged around 8 millibars, or about 0.8 percent of the atmospheric pressure on the Earth's surface. The polar caps seemed to be sheets of solid carbon dioxide (dry ice), while the thin atmosphere was mainly carbon dioxide with a trace of water vapor. The next Mars spacecraft, Mariner IX, arrived at Mars on November 14, 1971, and photographed 70 percent of its surface.

Then the Vikings were sent and, for the first time, American spacecraft landed on the red planet. Viking 1 was launched on August 20, 1975, and went into orbit around Mars on June 19, 1976. Its lander was separated and landed on Mars on July 20, 1976, while the orbiter continued to circle Mars and observe it. Viking 2, its twin, was launched on September 9, 1975, and Viking Lander 2 landed on September 3, 1976. What have these four spacecraft reported from Mars?

What emerges from thousands of photographs and miles of computer tapes is a planet different from the Earth and the Moon yet sharing some of the characteristics of each. The atmosphere of Mars is not only about one-hundredth the density of the Earth's but its composition is sharply different, as shown in Table 9-1.

While Mars's atmosphere is made up of much the same elements as Earth's, their proportions diverge widely.[2] Over Mars, carbon dioxide is predominant and oxygen is only a trace, the reverse of the situation on Earth. Mars, too, is extremely dry compared to Earth. Many scientists believe that the water vapor on Mars may have been dissociated by ultraviolet solar radiation into oxygen and hydrogen. The hydrogen, being so light, would have escaped into space, and the oxygen would have combined with elements in the soil such as iron, giving Mars its red color. Analysis of the isotope ratios of oxygen and nitrogen in its atmosphere indicates that Mars in the past may have had a much denser atmosphere, which was also richer in nitrogen.[3]

The fact that argon has about the same concentration in the at-

TABLE 9-1. Composition of Atmospheres of Earth and Mars

	Earth	Mars
Carbon dioxide	0.03%	96.20%
Nitrogen	78.00%	2.60%
Argon-40	0.90%	1.60%
Oxygen	21.00%	0.15%

mospheres of Mars and the Earth may be significant. Argon is a noble gas—that is, it does not react chemically with other elements. Traces of other noble gases, krypton and xenon, have also been discovered. Argon, krypton, and xenon are also heavy gases. They are therefore excellent indicators of the original state of the Martian atmosphere billions of years ago, for they could not escape into space or combine with the surface minerals.

The presence of nitrogen, even though in low concentration, proves that all the main life elements are there. Does life exist on Mars? This is possible, but doubtful. For all their sophisticated instruments, Viking Landers 1 and 2 could not give a conclusive answer.[4] One point is certain: the findings were not influenced by contamination from Earth microorganisms. The life-detection instruments were manufactured with rigorous clean-room methods and, after they were assembled, were heated for 54 hours in dry nitrogen at 120°C (248°F) before being shipped to Kennedy Space Center, where they were installed in the Viking Lander in a clean room and sterilized.

The biology package in the Lander contained three instruments that ran experiments designed to detect the presence of life: gas exchange, labeled release, and pyrolytic release. In the gas-exchange experiment, a little nutrient medium was added to the soil sample, the sample was incubated, and any gases given off were measured. Oxygen was detected, but that could have been due to peroxides in the soil.

The labeled-release experiment deviated from the gas-exchange experiment mainly in incorporating radioactive carbon, carbon-14, in the nutrient chemicals. If Martian microorganisms were in the soil, they would assimilate the nutrients, and so their gaseous wastes would be radioactive. The experiment, therefore, would be specific. It worked all too well. Radioactive carbon dioxide showed up immediately after the nutrients were added to the soil sample. Clearly, the carbon dioxide came from the peroxides in the soil, not from the metabolic activity of microorganisms.

The pyrolytic-release experiment took the opposite approach. It would look for proof of synthesis of organic compounds, not decomposition. A xenon arc lamp lit up the little chamber in which a soil sample was kept, and carbon dioxide and carbon monoxide were admitted. The sample was incubated for five days, then the lamp was turned off, the air pumped out, and the soil sample analyzed for carbon. Seven of the nine tests were positive, but the amount of carbon found was extremely slight, enough for only about 1000 bacteria at most.

The evidence from Viking Lander biology experiments, then, is ambiguous. In addition, a very sensitive instrument in the Lander, a gas chromatograph-mass spectrometer, determined that if organic compounds existed in the Martian soil samples, they must have had a concentration of no more than one part per billion. Instruments like the GCMS have detected organic compounds in the soils of Antarctica, the most inhospitable site for life on Earth. Microbes may be present in the soil of Mars, but it is difficult to account for the extreme scarcity of their organic remains. Carbon appears to be even scarcer in Martian soil than in lunar soil.

Life, then, may exist in the soil of Mars, but that is improbable. At both Lander sites, the soil did not get above freezing and was much colder a few centimeters down. The synthesis of traces of organic compounds is not necessarily proof of life, for they could be produced by purely chemical reactions in a lifeless Martian soil. The best way to clear up this complicated enigma would be to dispatch a scientific expedition to Mars, equipped with a sophisticated laboratory.

A determination that Mars is lifeless would make the planet easier to explore. There need then be no worry about bringing back alien life forms to Earth. Spacecraft to Mars would no longer have to be sterilized, which would reduce their cost and improve their reliability.

NASA Plans for Manned Expedition to Mars

NASA has been considering the manned exploration of Mars since the spring of 1962, when the Future Projects Office of the Marshall Space Flight Center initiated studies of possible Mars missions. Marshall was soon joined by other NASA centers. The first program, named EMPIRE (Early Manned Planetary-Interplanetary Roundtrip Expeditions), concentrated on plans for flybys of Mars and Venus in the early 1970s. Study contracts were let to Aeroneutronics, Lockheed, General Dynamics, and Douglas.

EMPIRE would depend upon NOVA for lifting it into orbit. NOVA was to be a gigantic four-stage rocket, about 152 meters (500 feet) high and 30 meters (100 feet) in diameter. It would have a mass of 15.9 million kilograms (35 million pounds) on the pad and be able to carry a 450,000 kilogram (one million-pound) payload. The first-stage engines would generate 200 million newtons (45 million pounds) of thrust. NOVA would carry a 180,000-kilogram (400,000-pound) spaceship into parking orbit. After checking the ship and determining the precise transmartian takeoff point, the six-astronaut crew would restart the engine and head for Mars. The voyage to Mars would take 613 days; then the ship would swing around the planet, head for Venus, fly past Venus, and return to Earth, where it would enter Earth orbit. The crew would enter an Earth Return Vehicle, detach from the spaceship, and return to Earth in it. The ERV would have some aerodynamic lift so that it could be flown to a landing.

The EMPIRE study never went any further, for NOVA was discarded as too expensive and replaced by Saturn. EMPIRE was succeeded by MEM (Mars Excursion Module study), assigned to Aeroneutronics; MMM (Mars Mission Module), assigned to North American Aviation; and ERM (Earth Return Module), assigned to Lockheed. MMM was to take the astronauts from Earth orbit to Mars orbit and, after they had explored Mars, back to Earth orbit. MEM was to take them down from Mars orbit to land on Mars and then bring them back up when they had finished their exploring stint. ERM would take the astronauts back to Earth from Earth orbit. Marshall Space Flight Center would study putting the pieces together.

From their start, plans for manned planetary expeditions had to compete with unmanned spacecraft missions to the planets. Since the unmanned spacecraft would be much smaller, lighter, and simpler, they would be cheaper. Whether they would be more effective was not considered.

When the MMM study was completed in 1966, NASA was concentrating its efforts on the Apollo project, and prospects for a manned Mars expedition were dim. After the triumphant landing of Apollo 11 on the Moon on July 20, 1969, interest in exploring the planets revived. In anticipation of the first lunar landing, President Nixon on February 13, 1969, set up a Space Task Group headed by Vice-President Agnew, with the NASA Administrator, the Secretary of Defense, and the Science Adviser to the President as members. The group was charged with preparing a program on the direction that the United States should take in space in the post-Apollo period. The Space Task Group was to report to the President by September 1, 1969.[5]

On the key issue of a manned Mars expedition, the Space Task Group wavered, taking refuge in safe generalities:

The Space Task Group sees acceptance of the long-term goal of manned planetary exploration as an important part of the future agenda for this

> *Nation in space. The time for decision on the development of equipment peculiar to manned missions to Mars will depend upon the level of support, in a budget sense, that is committed to the space program.*

The last sentence is so obvious it is banal. The Space Task Group continued:

> *Thus, the understanding that we are ultimately going to explore the planets with man provides a shaping function for the post-Apollo space program. However, in a balanced program containing other goals and objectives, this focus should not assume over-riding priority and cause sacrifice of other important activity in time of severe budget constraints. Flexibility in program content and options for decision on the specific date for a manned Mars mission are inherent in this understanding.*

Translated into English, this means: "Dump the manned Mars project." Agnew and his associates were keenly aware of the growing hostility to science in Congress, and they were not willing to urge Nixon to stand up to Congress by advocating a manned Mars expedition.

Werner von Braun, then Director of the Marshall Space Flight Center, described a plan for a manned Mars expedition at a hearing of the Senate Committee on Aeronautical and Space Science on August 5, 1969.[6] It was a major advance on previous plans.

Since, von Braun said, a realistic analysis required selecting dates for departure and return, the plan specified the spaceships would leave Earth orbit on November 12, 1981. Spaceships, not spaceship. To ensure safety, two identical spaceships, each with a six-man crew, would make the journey. Each would have a mass of 727,000 kilograms (or 1,600,000 pounds) after assembly in Earth orbit. Each spaceship would be 82 meters (270 feet) long and 10 meters (33 feet) wide. (Saturn/Apollo was 111 meters [365 feet] long and 10 meters [33 feet] in diameter.) Two giant nuclear stages (a nuclear engine with its tank for liquid hydrogen) would be strapped to the spaceship. The spaceship would also have its own engine. These 445,000-newton (100,000-pound) thrust engines would be developed through NERVA.

The diameter of the nuclear booster stages and that of the spaceship, 10 meters (33 feet), is not a coincidence. It is the diameter of the Saturn V. The spaceship and its nuclear boosters would be launched into Earth orbit in four flights by Saturn V stages. Once in Earth orbit, the sections would be assembled to form spaceships. Fully assembled and checked out in orbit, the Mars spaceships would be ready for their voyage.

In low Earth orbit of 555 kilometers (345 miles) altitude, the ships would be moving at 7.58 kilometers (4.71 miles) a second, only 3.6 kilometers (2.24 miles) a second less than Earth escape velocity. The ships would fire their nuclear boosters, accelerating to escape speed and into interplanetary trajectory. After the ships had reached their planned velocity, the nuclear boosters would be separated, turned around, and fired again. The boosters would go into elliptical orbit around the Earth and swing out as far as the Moon's orbit. When they reached perigee, their engines would be fired again, switching them into Earth orbit. Here, in effect, they would be left in storage to await their next mission. The uranium-235 in their reactors would have hardly been used. With a fresh supply of liquid hydrogen brought up from the Earth, they would be ready for their next voyage, which could be to the Moon, to Mars again, or elsewhere in the solar system.

Meanwhile, the spaceships would be traveling to Mars, which they would reach 270 days after leaving Earth orbit. After separating from its two nuclear boosters, each ship would be down to 295,000 kilograms (650,000 pounds). They would swing around the planet and

Figure 9-1. Earth departure—Von Braun's proposed Mars expedition. Von Braun proposed that two spaceships should be sent to Mars for safety's sake. Each spaceship has two nuclear boosters. (Courtesy of NASA, Marshall Space Flight Center)

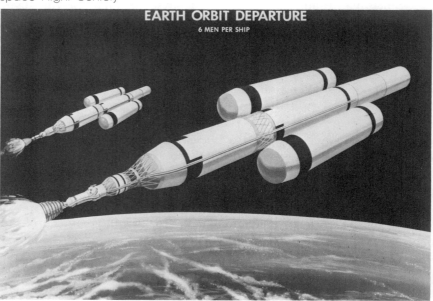

Figure 9-2. Mission weight history—Von Braun's proposed Mars expedition. Note that the weight of the spaceship returning to Earth orbit is a considerable fraction of its weight on departure for Mars. This shows one advantage of combining departure from Earth orbit with nuclear engines. (Courtesy of NASA, Marshall Space Flight Center)

fire their engines at the closest point to lower their velocity and revolve around Mars in a 24-hour orbit. The crews would send down unmanned spacecraft to the surface to take soil samples and fly them back to the ships, where they would be analyzed to determine their chemistry and whether they contained living forms.

That settled, three men from each ship would board the excursion module and fly down to Mars, landing by firing retrorockets. The excursion module would have an ascent stage, a laboratory, and living quarters. The men would explore the region around their landing point, collect rocks and soil samples, measure magnetic and gravitational fields, set up seismographs, take pictures, search for water, and so on. They would also experiment with growing plants in their laboratory.

Their 80-day stay coming to an end, the astronauts would take off in the ascent stage of the excursion module and ascend to rendezvous with the mother ships in Mars orbit. The two ships would now have a mass of 172,000 kilograms (380,000 pounds) each. The ships' engines would fire, taking them out of Mars orbit and on a heading toward Venus. On their way, they would cross the Earth's orbit. One

Figure 9-3. Landing on Mars—Von Braun's proposed Mars expedition. The manned lander is an improved version of the Apollo landing module. Also, since Mars has a thin atmosphere, the Mars explorers will have to wear spacesuits. (Courtesy of NASA, Marshall Space Flight Center)

hundred and twenty-three days after leaving Mars, they would be circling Venus. The crews would send two unmanned probes into the hot, thick Venusian atmosphere to conduct a radar survey of its surface. After 167 more days, the ships would reach Earth orbit on August 14, 1983. There, they would leave the ships in orbit and take another module to land on Earth.

A listing of all the modules needed for this complicated project is revealing.

Core ship (Mars Mission Module or Mother Spaceship)	2
Nuclear Strap-on Booster	4
Mars Excursion Module	2
Unmanned Mars Lander	4
Unmanned Venus Probe	4
Saturn V Stage	8
S-2 Stage	8
Earth Landing Module	2
TOTAL	34

At the time, a Mars expedition was held to be extravagantly expensive. The cost was often stated as $100 billion, a dubious estimate. Despite inflation, payload costs are declining and will continue to do so as space technology progresses. Since von Braun outlined his project to the Senate Committee, many advances have been made. Even now, an expedition to Mars could be much simpler and therefore cheaper than von Braun's complicated plan.

The Space Transportation System is logically the starting point for a solar system transportation network. Consider what the Space Shuttle could do to reduce the expense of a manned Mars mission. But first, let us assume that only one Mars ship is to be dispatched, not two, cutting the cost in half. Von Braun suggested having two spaceships to insure safety, but one should be so reliable that the safety of the crew would be assured.

In 25 flights, the Space Shuttles could carry 29,500-kilogram (65,000-pound) sections of the Mars ship to low Earth orbit for assembly. With five Space Shuttles in operation, each would have to make five cargo flights. With a turnaround time of two weeks, the Mars ship could be assembled in orbit in about three months. If, by the time the Mars Project is under way, SSTOs with a cargo capacity of 113,500 kilograms (250,000 pounds) are in service, only seven SSTO flights into orbit would be enough to take all the sections up for joining together.

The space station would facilitate construction of the Mars ship. Sections could be joined together while docked to the station. Assembly, checking, and outfitting would be much easier than working in open space, since a pressurized atmosphere, electricity, test equipment, and supplies would be easily available. Probably the job would go much faster.

Let us review the list of Mars modules of the von Braun project. The Space Shuttles are common carriers and would not, per se, be counted as part of the project. The Saturn V and S-2 stages would be unnecessary, since the Space Shuttle would take over their functions. The unmanned Mars Landers have been superseded by the Viking Landers. The nuclear boosters are required only if the NERVA engines are used. Replace them with the more powerful and efficient gas-core engine, and the Mars ship can perform its mission without them. Since the Mars mission will go only to Mars and back, the unmanned Venus probes can be left out. From 34 modules, we have come down to two—the Mars Mission Module (Core Ship or Mother Ship) and the Mars Excursion Module.

Von Braun was farsighted. He said in his testimony:

These reuseable vehicles can be used in the future not only to help us boost a module to Mars but also for supply and logistics; for example, to the Moon, if we want to run a research station on the Moon . . . . It will be a workhorse for all these high-speed missions that can be used over and over again. It must be refueled in orbit with another reuseable vehicle [the Space Shuttle] *that I will mention a little later.*

Von Braun had chosen NERVA solid-core nuclear engines for his Mars spaceships. They would have a specific impulse of about 850 seconds and a thrust of 333,750 newtons (75,000 pounds). Since 1969, however, the gaseous-core engine has appeared more and more promising. Robert G. Ragsdale of the NASA Lewis Research Center has analyzed a theoretical Mars flight with a spaceship propelled by a gaseous-core engine.[7] His conclusions are highly favorable.

Again, the Mars ship starts from Earth orbit. It has a mass of 2 million kilograms (4.5 million pounds), which includes a liquid hydrogen propellant load of 1.54 million kilograms (3.4 million pounds). The ship's five-person crew lives in its command module. The engine runs at 222,000 newtons (50,000 pounds) of thrust for 2½ days to attain 30 kilometers (19 miles) a second on its transmartian course. During this time, the ship accelerates at only 0.01 g. Thirty days after leaving Earth orbit, the ship starts its retrofire for insertion in orbit around Mars. The ship orbits Mars for several days, then heads back to Earth for a splashdown landing in the Pacific.

Ragsdale said that to make such a voyage a reality, the Space Shuttle would have to be on hand, the Skylab would have to demonstrate that humans can live in space for 60 days, a gas-core engine would have to be developed, and, most important, the United States would have to commit itself to a manned Mars mission.

The first two conditions have been met. The third and fourth are linked, for revival of the Rover Project would certainly have a manned Mars voyage as one of its aims.

One phase of the voyage is obsolete. After Viking, it became inconceivable that a manned expedition to Mars would not land and explore at least a small area of the planet. Ragsdale noted that by extending the trip to 160 days, the crew could leave their ship in a Mars parking orbit, descend with a 150,000-kilogram (333,000-pound) payload to the surface, stay there for 40 days, and then return to Earth. For this modification of the mission, the starting mass of the

Mars ship in Earth orbit would remain 2 million kilograms (4.5 million pounds). Ragsdale concluded with the important point that for fast planet missions, only the gas-core engine would do.

Present Capabilities for a Mars Expedition

Could the cost of a Mars mission be further reduced? Definitely, if the first mission is to be the beginning of the exploration and settlement of Mars. A space station in orbit around Mars would complement the Earth space station and help reduce the overall velocity requirement, the major influence on the cost of space flight. Fortunately, there is no need to construct a Martian space station. Nature has provided one free. It is called Phobos.

Phobos, the inner moon of Mars, revolves around Mars at 5980 kilometers (3715 miles) altitude in an almost perfectly circular orbit. Phobos is a potato-shaped object only 27 kilometers (17 miles) long with negligible gravity. Also, as Phobos revolves around Mars, one side always faces the planet, much as the near side of the Moon always faces the Earth. Its orbital velocity is 2.1 kilometers (1.3 miles) a second. As the escape velocity of Mars is 5.1 kilometers (3.2 miles) a second, a spaceship leaving from Phobos would need to add at most only 3.0 kilometers (1.9 miles) a second to escape from Mars.

If a small space station were built on Phobos, the Martian spaceship could dock there and transfer supplies and passengers to the Martian space shuttle, which would serve as a ferry between the Phobos station and the Martian base. Equipped with a nuclear reactor, the base could supply the spaceship with hydrogen for its return voyage to Earth. The ship would then have to carry only half as much hydrogen as it did for its first mission to Mars. The hydrogen would come from dissociating water ice from the north polar cap at a high temperature. The oxygen so derived would be invaluable for enclosed domes for Martian habitats.

Mars has a topography that appeals to the adventurous. The northern hemisphere is a vast plain with four enormous extinct volcanoes, the relics of gigantic eruptions. Olympus Mons, the largest volcano on Mars and, possibly, in the whole solar system, is 22,500 meters (75,000 feet) high and 600 kilometers (372 miles) wide at the base. The walls of the Olympus Mons crater are up to 2700 meters

(9000 feet) high and have slopes of 32 degrees. These volcanoes would challenge any mountain climber.

The southern hemisphere is heavily cratered and more moonlike in aspect. Besides its volcanoes, Mars has deep canyons, huge craters, winding valleys, high plateaus, and two polar ice caps. Valle Marineris near the equator is over 4800 kilometers (3000 miles) long, 120 to 135 kilometers (5 to 90 miles) wide, and over 6000 meters (20,000 feet) deep. Imagine a super-Grand Canyon that ran from San Francisco to New York and you will have some idea of the size of Valle Marineris.

The permanent north polar cap, made of water ice, is about 960 kilometers (600 miles) across and possibly over 300 meters (1000 feet) thick. It is much larger than the south polar cap, possibly because perihelion (closest approach to the Sun) comes during summer in the southern hemisphere. Thin sheets of dry ice, frozen carbon dioxide, cover the polar caps and wax and wane with the long Martian seasons. The permanent caps are more persistent. The presence of water ice, together with long, winding channels, streamlined "river bed" islands in dry, flat-bottomed valleys, and other features lead many scientists to believe that long ago Mars had a much denser, wetter, and warmer atmosphere along with surface water and even floods at times.

There are some indications then that ages ago, Mars was both wetter and warmer than it is now. Where did the water go? Some geophysicists suppose the water vapor was broken down by the solar ultraviolet into hydrogen and oxygen. The light hydrogen escaped from Mars into outer space, while the heavier oxygen oxidized the minerals on the Martian surface. There is another possibility. The water may have gone underground, in which case Mars may have retained much, if not most, of its primitive water supply. Michael H. Carr of the U. S. Geological Survey has suggested that as Mars cooled, the ground water sank into the porous soil and froze to form a thick layer of permafrost.[8] Meteor impact could have triggered thawing and outpouring of the water in fast floods.

Geologists scrutinizing the Viking Orbiter photographs have detected a great deal of evidence on underground ice. Charles Allen of the University of Oregon has found table mountains, moberg ridges, and pseudocraters.[9] Table mountains have flat tops and steep sides and are caused by volcanic eruptions under glacial ice. Moberg ridges are straight, narrow, steep ridges formed by eruptions underneath glaciers. Pseudocraters are small cinder cones caused by steam explosions. They are formed by lava flowing over water-soaked soil.

Even more exciting is evidence that ice deposits lie at shallow

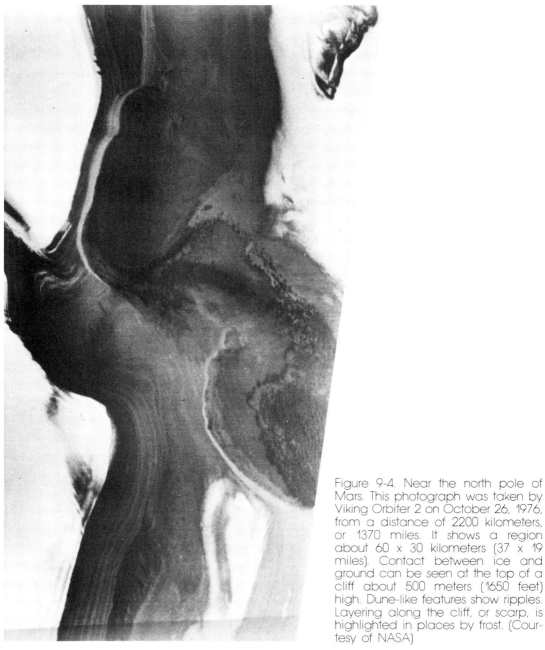

Figure 9-4. Near the north pole of Mars. This photograph was taken by Viking Orbiter 2 on October 26, 1976, from a distance of 2200 kilometers, or 1370 miles. It shows a region about 60 x 30 kilometers (37 x 19 miles). Contact between ice and ground can be seen at the top of a cliff about 500 meters (1650 feet) high. Dune-like features show ripples. Layering along the cliff, or scarp, is highlighted in places by frost. (Courtesy of NASA)

depths. The gound in Acidalia Planitia, wide plains in the northern hemisphere of Mars, is covered with cracks like those of permafrost soil in the Arctic areas of Alaska and Canada.

The issue of Mars exploration beyond Viking is often presented in terms of men versus machines. Oddly, the marvelous revelations of Viking through thousands of photographs and streams of data from its instruments have failed to arouse public interest in continuing the in-

Figure 9-5. A giant canyon on Mars. The western end of the huge canyon Valle Marineris, along with the equator of Mars. This is a mosaic of ten Viking Orbiter 1 photographs taken on August 22, 1976. In this area, a volcanic plateau is deeply dissected to form a complex series of interconnected depressions. At the edge of the plateau, numerous spurs jut out into the adjoining depression. These features may be produced by a combination of wind excavation and the movement of debris caused by alternate freezing and thawing of ground ice. (Courtesy of NASA)

Figure 9-6. A plain on Mars, as viewed by Viking Lander 2. The picture was taken on September 5, 1976, during the Martian afternoon. The porous boulder at the left edge of the picture is about two feet across. (Courtesy of NASA)

vestigation of Mars. Over the years, planetary exploration has come to mean sending unmanned spacecraft to the planets, with manned flights being relegated to the ever-receding future. The only plans seriously considered by NASA are to send another Lander with a roving automated vehicle or one that would bring back a small sample of Martian soil for laboratory analysis. It is no wonder that NASA's planetary programs fail to gain strong popular backing.

The combined cost of the two unmanned missions to Mars would certainly be over a billion dollars. They would return new information, of course—but could even a few ounces of Martian soil and rock analyzed in a laboratory in the Earth space station compare to what could be accomplished by humans exploring Mars for a month or more? A manned expedition may cost more to start, but its research yield/cost ratio would be far greater than that of any conceivable combination of instruments.

Sketch of a Mars Expedition

Let us consider what a manned Mars expedition might be like and, above all, what it could accomplish. We shall use some data from a study by Fischbach and Willis.[10]

The Mars spaceship is transported in sections to the space station by the space shuttles. The first section is docked to the station, and the other sections are joined to it in sequence. The engine chosen is an advanced space-radiator-gaseous-core nuclear reactor design. It has high specific impulse, over 5000 seconds, and high thrust-to-weight-of-engine ratio. After construction and outfitting is complete, the ship is thor-

oughly tested. Its fuel tanks are filled and it takes off with its eight-person crew on its long journey to Mars, a matter of some 40 days. Nearing Mars, the ship's engine fires to put it into a braking orbit and later to rendezvous with Phobos. Maneuvering gently with its reaction control jets, the ship comes to rest horizontally on Kepler Ridge.

First, the spaceship must be secured to prevent possible damage. Members of the crew drill holes deep into the rock around the ship, insert pipes, and fasten the ship to the ridge with heavy cables. Working on Phobos can be dangerous, since its gravity is trivial and drifting away is a constant hazard. The astronauts work in space suits that are virtually one-person spacecraft.

Their second task is to erect three antennas, one at the sub-Mars point that faces the planet, another at the sub-Mars point to face the Earth when it is in its field of view, and a third at the opposite end of Phobos to face the Earth when in its field of view. A small cabin is assembled in Hall Crater and connected to the three antennas by cable, so the astronauts can communicate with the Earth at all times. The Mars shuttle is unloaded and carefully set down. The four astronauts for the first descent enter and take their seats. The door is locked. Three astronauts push from one side, and the shuttle begins to drift away. At a safe distance, the shuttle pilot starts the engine, and the Mars shuttle is on its way.

The flight through the thin Martian atmosphere is partly aerodynamic, but landing is done with retrorockets. Like the Lunar Excursion Module, the shuttle descends vertically, landing at a high northern latitude about midway between the north polar cap and Olympus Mons. Phobos comes into view twice each day low on the southern horizon, so communication is easy. The crew does not immediately set out to explore. They plan to spend at least six months on Mars and there will be plenty of time for journeys. A flat spot is chosen for the site of the base. The crew clears away the rocks and smooths the powdery red soil. They carry out from the shuttle a long plastic bag and unfold it to its full 60-meter (200-foot) length. A compressed-air tank is attached, and soon the bag swells into a low, ellipsoidal, transparent dome. The dome is hermetically sealed with a tough plastic floor and is double-walled for insulation. Heated by the greenhouse effect of the sunlight, the dome soon becomes pleasantly warm. During the bitterly cold nights, space heaters will keep its temperature above 10°C (50°F). A plastic tunnel is run from the dome to the shuttle and the crew move in their effects and equipment. The spacious dome will be their living quarters, laboratory, workshop, storeroom, and communications post during their stay on Mars.

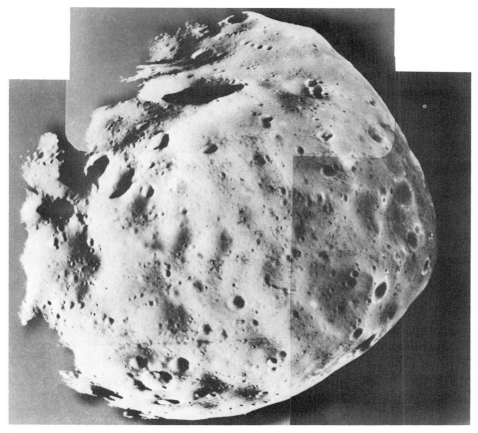

Figure 9-7. Phobos. This mosaic of pictures was taken by Viking Orbiter 1 on February 18, 1977, when it came as close as 480 kilometers (300 miles) to Phobos, the inner satellite of Mars. Mars is to the left. The large crater near the center top of the picture is Hall, 5 kilometers (3 miles) across. (Courtesy of NASA)

Several trips are made to Phobos to bring supplies. Next, the crew assemble a small Mars rover, which will serve both as a truck and a passenger vehicle. Its first trip is to a place behind a nearby hill, where the astronauts dig a pit for their small nuclear reactor. The reactor is divided into parts that can be handled by one person. After the reactor is completed, the fissile fuel is brought from the shuttle, heavily shielded, and inserted by remote control. A cable is run from the reactor to the base. The reactor will supply more than enough power for heating, lighting, the life support system, machinery, instruments, and other purposes. A large surplus is left for future growth demand.

The expedition has three primary aims: to learn as much as it can about the planet, to settle the question of life on Mars, and to begin

Figure 9-8. A manned expedition to Mars. This is an artist's concept of a manned expedition to Mars in the 1990s. Two manned capsules have landed together with rovers and a processing module. (Courtesy of NASA)

settlement. It also has a long-range aim on which it can only make a start: to transform Mars into a planet with a viable atmosphere.

This last aim, planetary engineering, is not idle fantasy. Scientists at the NASA Ames Research Center have conducted a serious study of ways to transform Mars into an Earthlike planet.[11] Dr. Robert D. McElroy coordinated the study, which brought together astronomers, chemists, biologists, meteorologists, geophysicists, and planetologists. After months of work, which included highly sophisticated mathematical analysis of the Martian climate and formulation of mathematical models of plant physiology, they concluded that Mars could be made over.

An atmosphere with enough oxygen for breathing and enough ozone to block deadly solar ultraviolet light could be generated by inducing plants to grow on the barren, bone-dry, and bitterly cold soil, absorb carbon dioxide and give off oxygen and water vapor. However, this process could take hundreds of thousands of years! How could it be speeded up? The planetary engineering of Mars then resolved itself into two approaches:

1. Melting one of the polar caps, which would add trillions of tons of water vapor to the atmosphere.
2. Creating new, Mars-adapted species of plants that would flourish and rapidly spread, despite the fiercely life-hostile Martian environment.

The north polar ice cap may contain 90,000 trillion kilograms (100 trillion tons) of ice. How could this enormous ice sheet be melted? Some scientists suggested spreading a layer of dark Martian soil over the ice. During the northern hemisphere summer, the soil would absorb the solar heat and gradually heat the underlying ice to the melting point. The ice would sublimate through the porous soil into the cold, dry air. (That is, it would evaporate without first turning into water.) As the ice cap very slowly shrank, the relative humidity of the air would rise; as a result, the temperature would go up, since water vapor is more effective than carbon dioxide in promoting the greenhouse effect.

This process, though, might take tens of thousands of years. Could plants do better? A survey of possible plants that could live and grow on Mars determined that only blue-green algae could survive the rigors of the Martian environment. They are anaerobic; that is, they do not require oxygen to live. However, their very survival on Mars is dubious, and new strains would have to be developed by genetic engineering. Special strains of "Mars" algae would have to be designed and created that would flourish in an environment harsher than the high dry valleys of Antarctica, enduring strong ultraviolet radiation, an atmosphere equivalent to that at 30,400 meters (100,000 feet) altitude on Earth, and temperatures barely above freezing for a few hours a day at best and plunging to lower than $-73°C$ ($-100°F$) at night.

Combining the two approaches might hasten the conversion of Mars into a life-supporting planet, but the process might take centuries.

A further possibility for transforming the atmosphere of Mars was not considered in the Ames study. Could nuclear energy melt the north polar ice cap? Nuclear reactors generate about 1000 megawatts or a billion watts, which could bring 4.2 million kilograms (4600 tons) of ice to the melting point each hour. On Mars, liquid water would immediately evaporate, even at 0°C. The average reactor could therefore evaporate about 35.3 billion kilograms (40 million tons) of ice each year, assuming that all the heat it produced was used for this purpose. However, a single reactor would take 2.5 million years to melt the north polar ice cap! To reduce this period to 10 years, 250,000 times as much power would be required. Even if all the nuclear power generated in the United States were applied, about 2500 years would pass before the cap was completely melted. These figures are, of course, approximate, for we still know very little about Mars. It is clear that planetary engineering will demand far more power than twentieth-century technology generates.

The first excursion of the explorers may be a short walk to take rock and soil samples, which will be examined for the possible presence of living microorganisms. The scientists will probably have on hand every type of analytical instrument used in life science research, far more than the four in the Viking Landers. They will be miniaturized but just as sensitive and reliable as their full-sized versions. The laboratory's microscopes, including an electron scanning microscope for morphology, may be critical in determining whether a tiny microscopic object is a living form or merely an inclusion in the soil. The search for life will take in looking for fossils as well as living organisms.

The Mars expedition will make a significant advance beyond the level of the Apollo Project. The astronauts who landed on the Moon acted as observers and collectors. They did no research there, for they had neither the time nor the facilities. The Apollo landings were only a brief reconnaissance. The manned Mars mission dare not be a reconnaissance. To send, let us say, six people to land on Mars and stay there no more than four days would be extravagant. The longer they remain, the more they can accomplish. From its inception, the Mars base should be designed as the nucleus of a permanent installation. Once it is self-sustaining, it should be permanently occupied.

Could that overriding question, the question of life on Mars, be settled in six months research at the Mars base? One way would be by looking for it. The Viking Landers did not have microscopes. Even if they had been so equipped, what use would they have been to the Mars investigators on Earth? How could the biologists tell whether a disk on the television screen showing a microscopic field was a living organism or a bit of mineral?[12] Life makes itself evident by its activity. Does it move? Does it increase in size—that is, grow? Does it multiply? Are its successors like their parents in shape and behavior?

Besides observing, the scientists could run numerous tests to answer the questions as they arose, something no instrument could do. It is astonishing that Mars, per the Landers, did not have any carbon or organic compounds in its soil, only carbon dioxide and a trace of carbon monoxide in the atmosphere.[13] The investigators would test the samples for organic compounds and also search for water in the samples, since life without liquid water is extremely unlikely. The Viking Lander instruments could not determine whether the soil samples it scooped up contained water. The scientists would also check for soluble salts of sodium, potassium, magnesium, calcium, chlorine, and phosphorus, which are good life indicators. A critical test would be comparing the gases which had been absorbed by the samples with the gases of the atmosphere.

Life is found everywhere on Earth from the summits of high mountains to the abyssal depths of the oceans. If life exists similarly on Mars, it should be present everywhere on the planet, but Martian life may be only marginal, existing only in life-favoring niches. Each exploring party would have as one of its primary objectives the obtaining of samples of sediments or soils from what may be such niches.

The first major journey might be to the north polar cap. The explorers would drive their rover over the curving frostfree strips that penetrate the cap like the spokes of a pinwheel. They would cross the ice on foot and drive the rover on bare ground. The drill they carried on the rover would be used for obtaining both ice and soil cores for later examination in the laboratory. The cores would be kept frozen in insulated containers. Back at the base, the cores would yield a rich store of data on the climatic history of Mars for possibly the last million years.

Once well into the ice cap, the explorers would send aloft a small, pilotless airplane under remote control. Flying at an altitude of less than 300 meters (a thousand feet) the airplane would take aerial photographs of the icy landscape of the polar cap. Then it would be guided back to the base where its roll of film would be developed.

The first exploring trip, like all the trips taken by the explorers, would have both scientific and practical purposes. The ice not reserved for scientific study would be melted, tested for chemical purity, and pumped into the base water tank.

The second exploring trip would be to Olympus Mons. Climbing that 22,800-meter (75,000-foot) mountain might be, in some respects, easier than the assault on Mt. Everest. The explorers would be wearing Mars suits, space suits modified for outdoor life on Mars, so the slight difference in atmospheric pressure between the base of the volcano and its summit would not be noticeable. The explorers would search for the easiest route, drive the rover up the slopes as far as it could go, and then hike to the rim of the vast caldera. The descent to the crater floor would be hazardous enough to satisfy the hardiest mountain climbers.

At the main base, meanwhile, the hundreds of samples piling up, the rolls of film and tape, the explorers' notebooks, and the laboratory data would keep the staff very busy. The findings would answer some questions and pose many more, for the world of Mars may be almost as complex as the world we live in. Besides searching for life, the investigators would prepare a classification of the Martian rocks into sedimentary, igneous, and metamorphic types and their sequence. The absolute age of the rock strata would be determined by a miniaturized accelerator-mass spectrometer that counts directly the

Figure 9-9a. Olympia Mons. This is an artist's sketch of Olympia Mons, a giant volcano on Mars. It is 23 kilometers (75,500 feet) high, and is the largest mountain in the Solar System. Its base is 600 kilometers (380 miles) across. (Courtesy of NASA)

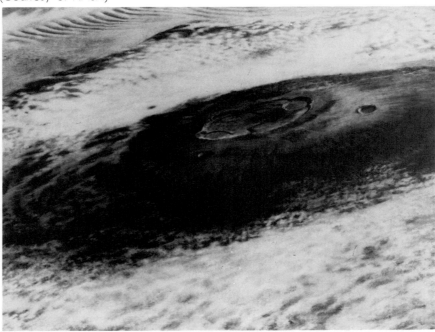

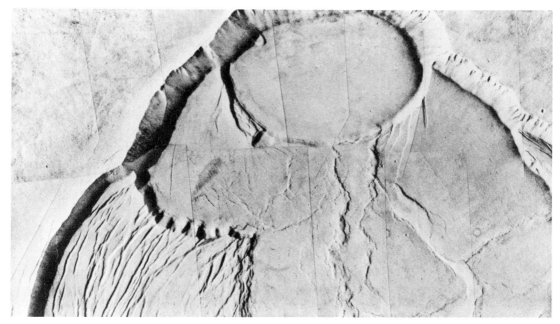

Figure 9-9b. Crater of Olympia Mons. Viking Orbiter 1 took a mosaic of photographs of part of the crater of Olympia Mons on June 13, 1976. The crater at top center is about 25 kilometers (15.5 miles) across. The crater walls are 2.4 to 2.8 kilometers (1.5 to 1.7 miles) deep with a slope of about 32 degrees. (Courtesy of NASA)

number of atoms given off by natural radioisotopes, the clocks of Nature.[14] There would be so much to do that as the planned six months drew to a close, the desire to extend their stay might become stronger and stronger. What better place to do research on Mars than on Mars itself?

With an abundant supply of water, a smoothly functioning life support system, and a nuclear reactor for assured power, remaining at the base for more than six months would not be difficult. Food plants could be cultivated by hydroponics and, if the explorers had foresightedly brought poultry with them, they could supplement their plant diet with animal protein. The spaceship on Phobos might start on its return voyage to Earth with tons of samples, a huge data file, and film by the kilometer, but with only part of its crew, the remainder having decided to stay on Mars until the spaceship came back. The Mars base would have become the Mars colony.

Notes

1. Ernst J. Öpik, "The Martian Surface," *Science*, 153:3735 (July 15, 1966), 255–265. See also E. Öpik, *Irish Astronomical Journal*, 1, 22 (March 1950).
2. S. Ichtiaque Rasool, Donald M. Hunten, and William M. Kaula, "What the Exploration of Mars Tells Us about Earth," *Physics Today*, July 1977, pp. 23–32.
3. Michael R. McElroy, Yuk L. Yung, and Alfred O. Nier, "Isotopic Composition of Nitrogen: Implications for the Past History of Mars' Atmosphere," *Science*, 194 (October 1, 1976), 70–72.
4. Harold P. Klein et al., "The Viking Biological Investigations: Preliminary Results," *Science*, 194 (October 1, 1976), 89–105.
5. *The Post-Apollo Space Program: Directions for the Future*, Space Task Group Report to the President, September 1969.
6. Statement of Dr. Werner von Braun, Director, George C. Marshall Space Flight Center, *Future NASA Space Programs*, Hearing of the Senate Aeronautics and Space Science Committee, August 5, 1969, pp. 14–30, 91st Congress, First Session.
7. Robert G. Ragsdale, "To Mars in 30 Days by Gas-Core Nuclear Rocket," *Astronautics and Aeronautics*, January 1972, pp. 65–71.
8. Michael H. Carr, "Formation of Martian Flood Features by Release of Water from Confined Aquifers," *Second International Colloquium on Mars, January 15–18, 1979*, NASA Conference Publication 2072.
9. Charles C. Allen, "Volcano/Ice Interaction on Mars," *Second International Colloquium on Mars, January 15–18, 1979*, NASA Conference Publication 2072.
10. Laurence H. Fishbach and Edward A. Willis, "Performance Potential of Gas-Core and Fusion Rockets: A Mission Application Survey," *Journal of Space-Craft and Rockets*, 9:8 (August 1972), 569–570.
11. M. M. Averner et al., "On The Habitability of Mars, an Approach to Planetary Ecosynthesis," NASA SP-414, 1976, NTIS: N77-12718.
12. *Post-Viking Biological Investigations of Mars*, Space Science Board, National Academy of Sciences (Washington, D.C., 1977).
13. *Strategy for Exploration of the Inner Planets: 1977–1987*, Space Science Board, National Academy of Sciences (Washington, D.C., 1978).
14. Richard R. Muller, "Radioisotope Dating with Accelerators," *Physics Today*, February 1979, pp. 23–30.

10
EXPLORING THE SOLAR SYSTEM

The space enterprise has created a diversified family of space vehicles: orbiters, probes, landers, satellites, and spaceships, each with a specific purpose. In the coming decades, mankind can build via these vessels a complete network that will span the solar system from the neighborhood of the Sun out to beyond Pluto. Space colonization will follow space exploration, and bases and colonies can be established throughout the solar system.

Understanding must be one of the major goals of space exploration. We take the constancy of the Sun for granted, even though it controls the climate, the oceans, the atmosphere, and the very existence of life on Earth. Long-range weather forecasting and the predicting of climatic trends will be impossible until the complex chain of events from the Sun's core to the surface of the Earth is fully monitored and thoroughly understood. Spacecraft have transformed our knowledge of the Sun, but we still do not have a full picture of its behavior. For one thing, our view of the Sun is a side view. We know little about what is happening around its poles.

This gap in our view of the Sun must be filled. NASA and ESA are working on two spacecraft to be launched into out-of-the-ecliptic missions. They will soar over and under the Sun's poles and back to Earth orbit. These first spacecraft to voyage far above and below the ecliptic plane of the solar system will discover how the solar wind and magnetic field of the Sun penetrate these regions of space. The twin spacecraft are forerunners of the solar surveillance network that will monitor the Sun's atmosphere.

There is only one way that the Sun's core can be observed directly: through neutrinos. Neutrinos have the astonishing property of being able to pass through very large amounts of matter without being stopped. Fortunately for science, a very tiny proportion are stopped by matter. Astronomers are trying to detect the neutrinos produced in the Sun's interior by fusion reactions. Almost all of the neutrinos generated by these reactions escape into space. Measuring the miniscule fraction of neutrinos that can be stopped will give the solar physicists important clues on the Sun's future energy output and the current physical conditions deep inside the Sun.

The solar neutrino detector now being used is a huge vat of cleaning fluid deep in a gold mine. The cosmic rays that permeate space are absorbed by overlying rock before they can reach the vat and confuse the readings of the attached instruments. The neutrino data, though, are puzzling, for apparently the Sun is producing fewer neutrinos than it should according to the best solar interior models. Additional data and new experiments followed by more theoretical work on the structure of the Sun are needed so that we can reach a satisfactory explanation.

The Sun must be ringed with satellites so that it can be continually observed. Three satellites would orbit the Sun in the plane of the solar equator at heliosynchronous distance, each observing the same face of the Sun and moving as if it were located at a fixed point over the surface, like the geosynchronous satellites that circle the Earth. Another three solar satellites would orbit the Sun in polar orbit, 120 degrees from each other, so the polar regions of the Sun would also be under continual surveillance. Each satellite would carry instruments that would detect and measure the full range of the Sun's particle and electromagnetic radiation output. They would communicate with each other continually. Those facing the Earth would relay data to the World Solar Observatory on Earth. Here, besides the solar physicists' research facilities, computers, terminals, and other equipment, there would be a hall for the public to witness this grand and eternal spectacle. The faces of the Sun as seen by these satellites would be projected on the black walls. The spectators would see the sunspots, the spicules, the prominences, the faint white corona, the coronal holes, and other aspects of the raging Sun as no human has ever before seen them.

The World Solar Observatory would also receive a flood of data from a solar system network of satellites, orbiting the Sun at distances from within the orbit of Mercury to far beyond Pluto. Some of these satellites' orbits would lie in the plane of the solar system while others would be inclined to it. They would measure the solar wind, the interplanetary magnetic field, solar cosmic rays, and cosmic rays coming from the Galaxy. Among their services would be giving early warning of solar flares and the particle streams they eject. These warnings would be relayed to spaceships, bases, and colonies throughout the solar system. This service would be important, for at the distance of the Earth from the Sun, the radiation and particles from a giant solar flare could kill an unprotected spaceship crew or unprotected astronauts on the Moon.

The major goal of manned space exploration will be the investigation of the bodies of the solar system, other than the Sun. Close-up monitoring of the Sun will probably be left to solar satellites, as described. Exploration and colonization of the Moon was discussed in Chapter 6 and of Mars in Chapter 9. The other bodies to be explored by humans will probably be first Mercury and the larger asteriods, with the outer planets and their satellites scheduled for later missions. Venus is a special case that probably will be observed at a distance for a long time before manned exploration is attempted. Bases would most likely be located on the largest satellite of each of the giant planets Saturn, Uranus, and Neptune. The exception would be Jupiter, since its largest satellite, Ganymede, orbits within its dangerous radiation belt. Pluto is a small planet and probably rocky, so humans should be able to land on it and set up a base. The probable base sites, then, for the outer planets and their moons are: Callisto for Jupiter, Titan for Saturn, Titania for Uranus, and Triton for Neptune.

A Mercury Base

Mercury is the nearest planet to the Sun and the second smallest. It has a weak gravitational field, and, due to its proximity to the Sun, its surface becomes very hot during the day. The combination has kept Mercury from having an appreciable atmosphere. Mercury may be

Figure 10-1. Mercury. Mercury's northern limb as taken by Mariner 10 on March 29, 1974, at a distance of 690 kilometers (420 miles) from the surface. A prominent east-facing scarp extends from the limb near the middle of the photograph southward for hundreds of kilometers. The bottom of the picture is 580 kilometers (360 miles) across. The "tear" in the edge is an artifact due to loss of data. (Courtesy of NASA)

similar in composition to the Earth, since its average density is almost identical. This suggests a rather high metal content, which should make it very interesting to geologists.

The days and years on Mercury are peculiar. Its rotational period is coupled to its period of revolution around the Sun. The little planet rotates on its axis with respect to the stars (its sidereal day) once in 58.6 Earth days. This is two-thirds of the time of 88 days that it takes to go once around the Sun.

The solar day is that time that elapses from when the Sun has risen to be at a selected point in the sky (any point will do) until it reaches that same point again. On Mercury, the solar day is 176 Earth days, or twice the length of the period of revolution around the Sun. Each hemisphere of the planet faces the Sun at alternating perihelia, when the planet and the Sun are closest. The lengths of the day and year on Mercury appear reversed. From the viewpoint of the exploration of Mercury, however, the days and nights would each be 88 Earth days in length. This situation on Mercury is unique in the solar system.

The base on Mercury would probably be located near the north or the south pole, where the Sun is always low in the sky and the daytime temperature is considerably lower than at the equator. The base would be within a crater for protection against the intense solar heat. Since Mercury has no atmosphere, the base could be shielded from the Sun by a large metallic awning that would be coated to reflect as much of the sunlight as possible and would be supported by metal heat pipes that would be sunk in the rock without touching the base. The awning could be cooled by a circulating liquid.

The scientists on duty at the base would usually operate the solar observing instruments by remote control. Night would be the preferred time for manned exploration and prospecting, since cold is much less a problem to planetary explorers that heat. To avoid temperature extremes, they would first explore regions near the terminator, the day-night line, where the surface temperatures are near 0°C to 100°C. The scientists would travel in completely enclosed, well-heated, comfortable rovers. In both equatorial and polar orbits, there would be three multipurpose satellites each that would enable them to keep in contact with the base and the Mercury space station at all times—a vital safety precaution. The base would have a large radio telescope that would serve both for radioastronomy studies of the Sun and for communication with the Mercury space station, the Earth, and the solar system communications network.

The Mercury space station would be the main link between the Mercury base and the Earth. Here, interplanetary spaceships would dock and transfer their cargo to shuttles, which would travel from the station to the base. The station would also be a second major solar observatory as well as an observing post for Mercury. Low orbiting satellites and maneuverable hovering rocket vehicles could map the surface in fine detail, besides making a wide variety of scientific measurements. Supplemental data could be provided by unmanned land rovers, which could also take soil samples. A grid of automatic land stations could be set up to monitor Mercury quakes, measure temperature changes, and observe the weak magnetic field of Mercury and its interaction with the solar-wind particles.

At first, the Mercury base would certainly be manned only by scientists, engineers, and technicians on tours of duty. Eventually it might develop into a colony where people would live permanently, and it might even have a hotel for tourists, if it proved to have unforeseen attractions. Very likely, Mercury will be an intriguing world to explore and one that will furnish important data to a score of sciences, yet it will be a more forbidding place than the Moon on which to live.

Observing Venus

Venus has proven to be a great disappointment to the space minded. How could a planet so like the Earth in size, in mass, and in distance from the sun be so different from the Earth? Is the difference due to its being closer to the Sun? This is one of the scientific questions about Venus that has to be answered. Without a specially designed "Venus" space suit, which would resemble a one-person submarine, an astronaut on the surface of Venus would be simultaneously broiled, crushed, and asphyxiated. Hence, Venus in its present state is unfit for human habitation, and visits to Venus would be very dangerous.

Venus rotates very slowly on its axis. It has a rotation period of 243 days with respect to the stars. Moreover, it rotates in a direction opposite to that in which it goes around the Sun. Its revolution period—the time it takes to orbit the Sun once—is 225 days. These two periods combine to give Venus a solar day of 127 Earth days. If Venus had a thin atmosphere like the Earth's, the difference in temperature between day and night would be very great, but in fact it has a very thick atmosphere, which distributes the solar heat efficiently from its daytime to its nighttime side.

Can Venus be transformed into an Earthlike planet? To do this, we must first understand how it came to reach its present state. This will require a long, thorough study of Venus, which should also provide us with a deeper understanding of how the atmosphere of another terrestrial planet functions and so could lead to deeper insight into the Earth's weather. The main Venus base of operations would be a manned space station that would orbit the planet every few hours. Satellites high above its equator and others circling in their polar orbits would watch its weather continually. Balloons floating at various altitudes in its dense atmosphere would telemeter data on temperature, winds, density changes, chemical composition, and atmospheric pressure. Automated rovers would crawl over its fiery terrain, sending back

Figure 10-2. Venus in ultraviolet light as viewed by Mariner 10 on February 10, 1974. The planet's clouds show a prominent swirl at the south pole. (Courtesy of NASA)

photographs and analyses of its soil and rocks. A network of stationary landers would report on atmospheric conditions at the surface and on Venus quakes, if they exist, and on heat flow. From time to time, specially equipped landers and rovers would launch rockets that would carry rock, soil, and atmospheric samples back to the space station for analysis. Over the years, the network could grow and become more sophisticated, becoming a parallel on a smaller scale of the Earth's meteorological and geophysical services. The detailed knowledge of the Venusian atmosphere flowing from this intensive research could in time lead to the possibility of terraforming the planet.

Transforming Venus would be a gigantic project. It would require reducing the atmospheric pressure 90 times, lowering the surface temperature from 427°C to 27°C, and speeding up its period of rotation.

Carl Sagan has proposed a plan for modifying the atmosphere of Venus.[1] In 1961, when spacecraft observation of Venus was just beginning, he suggested seeding the upper atmosphere of Venus with blue-green algae. The algae would convert the carbon dioxide into oxygen and so decrease the greenhouse effect. The atmosphere would become more transparent, radiate more heat into space, and slowly cool. As its temperature dropped, the atmosphere would become saturated with water vapor, which would start to rain and eventually form bodies of water on the surface. This would set up a terrestrial type of circulation. Possibly, the carbon from the carbon dioxide would eventually be deposited as layers of carbonates on the surface.

However, maintaining a terrestrial type of atmospheric circulation probably would require a rotation period of a few days, and so Venus would have to be spun up, which would require an enormous amount of energy. One possibility would be to deliberately crash an asteroid into Venus. An iron asteroid with a radius of 100 kilometers (62 miles) striking Venus on its equator at almost grazing incidence at a terminal velocity of 100 kilometers (62 miles) a second could speed up its rotation rate to about 150 days.

Visiting the Asteroids and Comets

The asteroid belt in which most of the asteroids or minor planets orbit is situated between Mars and Jupiter. Many of the asteroids are grouped in families.

Most of the asteroids in the asteroid belt are closer to Mars than to Jupiter. They may be the remains of what would have become a planet in the early stages of the formation of the solar system had it not been for the influence of Jupiter. Ceres, the largest asteroid, is 955 kilometers (592 miles) in diameter, and it probably contains about half of the mass of all the main-belt asteroids combined. Their total mass is probably less than that of the Moon. Only a fraction of the asteroids have been discovered so far, since they are mostly small and faint.

The few asteroids whose orbits cross the orbits of the inner planets, including the Earth's, are called Apollo asteroids. The extremely dark Trojan asteroids are locked in stable orbits at Jupiter's distance from the Sun, either 60 degrees ahead of or 60 degrees behind Jupiter as seen from the Sun.

There is also Chiron, an object of asteroidal size, 300 kilometers (190 miles) in diameter, which was recently found by Charles Kowal of the California Institute of Technology. Chiron orbits with a period of revolution of 51 years and moves from within the orbit of Saturn out to the distance of Uranus. It may be a member of a new class of hitherto undiscovered minor planets beyond the orbit of Saturn.

In recent years, interest in asteroids has greatly increased and our knowledge of them has been rapidly growing. It now appears that many asteroids may have tiny moons. After it was found that 532 Herculina possibly had a satellite orbiting it, several astronomers took up the search for more "minor satellites" and found other asteroids that seem to have their own satellites. Asteroids are difficult to observe, even with large telescopes, and are usually discovered as streaks on astronomical photographs (a streak shows that a body is moving against the background of the stars). Minor satellites are, of course, even harder to detect, so astronomers use photoelectric photometers, instruments that measure the intensity of light, to study the occultation of a star as an asteroid passes between the star and the observer. If the asteroid is moonless, then the light curve drops sharply as the asteroid occults the star and goes up as it passes away from the star. However, if the light curve drops twice, then the astronomer can assume that a second body occulted the star. But, for the second body to be a satellite, the secondary occultation should occur soon before or after the primary one. Further, since it is a solid body, it should block the starlight equally in all colors. Finally, several occultations need to be observed so the evidence can be seen for the satellite's orbital motion. This has to be done to confirm the existence of the possible Herculina 532 satellite and of other candidate asteroid satellites.

The asteroids may furnish clues to the origin and evolution of the solar system, be sources of raw materials, and serve as sites for the anchoring of instruments. The meteorites found on the Earth's surface appear to be fragments of asteroid material. One bit of evidence is that the orbits of some of them definitely had their aphelion points in the asteroid belt.

The asteroids can be divided into distinct classes. About 10 percent of the 200 largest asteroids have spectra that resemble those of stony iron meteorites, as determined in the laboratory. These asteroids have diameters in the 100- to 200-kilometer (60- to 120-mile) range and are found mostly in the inner area of the asteroid belt closest to Mars. They may be parts of the metallic cores of larger minor planets that were sufficiently large to melt internally and so rearrange their chemical composition. The remnants are the results of collisions that broke off the overlying silicate layer.

Some 80 percent of the asteroids have spectra that resemble the very primitive carbonaceous chondrite meteorites. They contain minerals characteristic of the solar nebula and apparently contain water bound chemically to various compounds. These asteroids may never have been fragmented by collisions or were never big enough to become heated internally, and so they retained their original surfaces. Some members of this class may be surface fragments from larger bodies.

As part of the program of solar system exploration, planetary scientist will want to detect all the Apollo class asteroids and to determine their orbits, both to keep track of these navigation hazards and to select those to visit. If any are detected that will eventually be on a collision course with the Earth, it should be possible to visit them and to make minor course corrections by various propulsion devices so that a catastrophe will be avoided. They could be maneuvered into orbit around the Earth or about another inner planet.

Advocates of space colonies, particularly Dr. Gerard K. O'Neill, have suggested that the asteroids could be mined for their minerals. Since the cost of transporting material from an asteroid may well be less than that of bringing it up from the Earth, due to the asteroid's lesser gravitational attraction, this is an intriguing idea. A few asteroids may be almost entirely iron and nickel, while others may contain great amounts of these metals and still others may have much water locked up in their surface rocks. If these substances cannot be easily obtained from the surfaces of planets and moons, then they will likely be extracted from asteroids. However, this will be done only if it is more economical than extraction from the Earth or the Moon. The development of the fusion torch may make every solid body in the solar system an economical source of raw material.

Probably, the first asteroids to be visited will be Earth orbit-crossing Apollos that appear to be composed of material similar to L-type chondrites, a type of stony meteorite. These first missions would make scientific investigations, including taking soil and rock samples and setting up observing stations for monitoring the interplanetary medium. While their orbits might not be as suitable as those of man-made satellites, their large surface areas and the use of their surfaces as a foundation makes them attractive. The instruments could be fastened directly to the surface rock. This minimizes the framework needed. A greater part of the mass left on the asteroid could be devoted to scientific experiments than for a man-made satellite, since the asteroid replaces much of the satellite's structure.

An attempt would be made as soon as practical to visit various types of asteroids, especially the dark asteroids and the ones that appear to be made up mostly of nickel and iron. A base for exploring the asteroid belt might be established on the largest asteroid, Ceres, which has a diameter of 955 kilometers (593 miles). Ceres is almost twice as large as the next largest asteroids, Pallas and Vesta.

From beyond the orbit of Pluto out to about a light-year from the Sun, there are perhaps a hundred million comets in the Öort cloud with a total mass of, at most, a few tens of Earth masses. Once in a while a comet has its orbit changed so it can closely approach the Sun. If it comes too near to a planet, its orbit may be changed to an ellipse having its aphelion within the orbit of Pluto. It is then called a periodic comet.

Since comets contain matter that probably is unchanged since the formation of the solar system some 4.6 billion years ago, we will want to send missions to some of them to obtain samples. They should provide important clues on the condition of the solar nebula. Also, although cometary nuclei are probably only about 1 to 10 kilometers (0.6 to 6 miles) in diameter, their surface areas are large enough for use as platforms for scientific experiments. A rendezvous mission with a comet would sample its surface material and set up an unmanned scientific base on the nucleus. Comets that had passed perihelion with the Sun would be preferred. This assumes, of course, that the nucleus is sufficiently solid to support the base—a likely assumption at the Earth's distance from the Sun. Placing a transmitter at the base would enable scientists to keep accurate track of its position. Nonperiodic comets with aphelia far beyond the orbit of Pluto may become mankind's first deep space probes.

The four Galilean satellites of Jupiter are sufficiently large and massive to be used as major bases for exploring the outer solar system. Their density decreases with their distance from Jupiter—from 3.5 times that of water for Io, the innermost Galilean satellite, to 1.6 times for Callisto, the outermost. The ratio of rock to ice decreases correspondingly.

Io is a colorful reddish-brown globe with a bright yellowish-orange band along its equator and dark reddish sulfur deposits at its poles. The surface seems to be overlaid with deposits of evaporated salt crystals. Io generates radio noise as its passage disturbs Jupiter's magnetic field. Voyager I revealed that Io has active volcanoes whose spectacular eruptions send clouds 300 kilometers (186 miles) high at speeds of over 3200 kilometers (2000 miles) per hour. Not only is Io the only body in the solar system, other than the Earth, discovered to have active volcanoes, but it has more than a hundred of them, 25 kilometers (15 miles) or more across.[2] Bigger than the Moon, with a diameter

Figure 10-3. Jupiter and two of its satellites. Voyager 1 took this photograph on February 13, 1979. Io is above the Great Red Spot while Europa is seen against the clouds of Jupiter. (Courtesy of NASA)

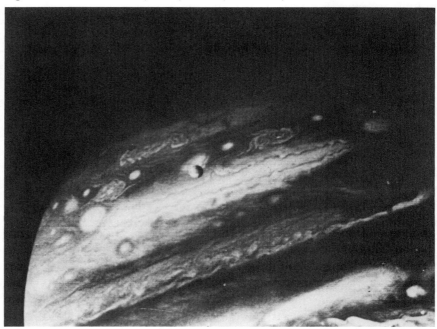

of 3640 km (2660 miles), Io orbits Jupiter in a little more than two days at the dangerously close distance of 420,000 kilometers (260,000 miles). Io is immersed in a glowing cloud of yellow, ionized sodium. This moon is strange, colorful, and dangerous—to be explored only with great caution.

Europa, in orbit about 670,000 kilometers (415,000 miles) from Jupiter, is the smallest of the Galilean satellites. It is a little smaller than the Moon, with a diameter of 3500 km (1900 miles). Europa is crisscrossed with faults hundreds of kilometers in length and appears to have irregular patches of ice. Perhaps this moon has huge glaciers.

Ganymede is a very large satellite; its diameter of 5200 kilometers (3200 miles) makes it larger than the Moon or even Mercury. If it orbited the Sun instead of Jupiter, it would be recognized as a planet. As it is, it moves around Jupiter at a distance of about a million kilometers (650,000 miles). Ganymede has a density only half that of the Moon, so it may be a mixture of rock and ice. Near its equator is a large dark area with light streaks, which may be a crater.

Unfortunately for explorers, Jupiter has a vast radiation belt that may extend as far out as 1.4 million kilometers (881,000 miles) from the giant planet.[3] This "Trapped Radiation Belt" is composed of streams of protons and electrons, along with other nuclear particles, circling Jupiter under the guidance of its magnetic field much as the Van Allen Belts circle the Earth. But the Trapped Radiation Belt is far more intense and dangerous than the Van Allen Belts. A spaceship passing through the most intense region of the Trapped Radiation Belt would receive a dose of about 3 million rads. The radiation dose inside the ship, unless it were shielded, would be about 300,000 rads. Some bacteria would survive, but no higher organisms would. Humans and mammals could not withstand this superlethal dose.

Jupiter's radiation belt, however, has been only penetrated, not fully mapped. It may vary in extent and intensity with time, and possibly it is far less intense farther out than it is near its surface. In any event, Callisto is the best candidate for a Jovian base, since it is the most distant of the Galilean moons from Jupiter.

Callisto is 1.9 million kilometers (1.5 million miles) from Jupiter and possibly beyond the radiation zone. Its surface is mottled with bright and dark spots. The light spots resemble the Moon's rayed craters. Like the Moon and the other Galilean satellites, Callisto always shows the same face to its central planet. The base will always have Jupiter in view if it is located on the near-side hemisphere, as it surely would be. Callisto, with its diameter of 5000 kilometers (3100 miles), is slightly larger than Mercury but its mass is only a third as much, so its surface gravity is only about one-tenth of the Earth's. It is very cold, with a surface temperature of about 105K (−270°F) and also very dark, the darkest of the Galilean satellites, which probably means that it has much bare rock exposed.

Callisto revolves about Jupiter in an almost circular orbit with a sidereal period of 7.155 days. Since most of the satellites rotate on their axes in the same time that they revolve around their planets, the solar day on Callisto is about 7.4 days. The Sun appears only 6 minutes of arc across, the brightest star in the black sky with a disk just discernible to the naked eye. Even so, it would give about 14,000 times as much light as the full Moon does to the Earth. Sunlight reflected from Jupiter would also light up the terrain.

Humans would need protection to live on this frigid world. We

Figure 10-4. Callisto, as seen from Voyager 1 on March 6, 1979, when the spacecraft was 350,000 kilometers (220,000 miles) from Callisto. Features as small as 7 kilometers (4.2 miles) across can be seen. Callisto's surface is made up of water-rich rock frozen at 120K (-243°F). The prominent bulls-eye-type feature is a large impact basin similar to Mare Orientale on the Moon and the Caloris Basin on Mercury. (Courtesy of NASA)

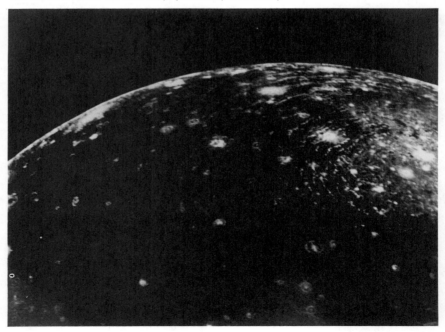

can imagine large, low domes that would be islands of light, warmth, and life in this frozen wilderness. The Callisto base would serve as the center for the exploration of Callisto and the other 13 moons of Jupiter as well as Jupiter itself. A small space station in polar orbit might serve as a transfer point for cargo. Orbiting satellites would map the surfaces of all the Jovian moons in fine detail and make geological surveys. Rovers could return soil and rock samples, furnish surface views, and set up automatic stations. Manned visits would investigate promising areas for metals and minerals, study the geology of the terrain, and attack other specific problems. Since Callisto is the most habitable of the Jovian moons, much effort would be devoted to its exploration with manned rovers. The ice sheets should contain more than enough water to sustain life at the Callisto base, which could expand into a network of colonies.

Jupiter would have to be explored by unmanned satellites. A synchronous satellite network would be set up with satellites both in equatorial and polar orbit. Balloon-borne instruments would float in the vast atmosphere, while sampling probes would descend into it to bring back samples of the gases and perhaps even samples from the deeper layer where it begins to become a slush.

Since Callisto has chemicals required to sustain life, there could be an attempt to terraform it. Krafft Ehricke of the Space Global Company has proposed artificial suns that he calls "helioids" for the moons of the outer solar system.[4] The helioid would orbit the moon and bathe it in light and warmth. It would appear as a shining round disk or point in the dark sky, too bright to look at. The helioid could be placed in an orbit far enough above the surface of the moon to give it a day of reasonable length. For Callisto, the orbital radius of the helioid would be 100,000 kilometers (62,000 miles) to give the moon a 24-hour day. To complete the transformation to a terrestrial-type environment, an artificial atmosphere would have to be provided. Since enormous energy would be required to establish helioids and an artificial atmosphere, it is likely that, at first, areas under domes would be terraformed, using artificial light. The helioid-atmosphere project would be undertaken only when the progress of science and technology made it readily feasible.

The plan of planetary investigation would be similar for the three outer Jovian planets, Saturn, Uranus, and Neptune: a network of satellites to monitor the planetary weather, atmospheric balloons, atmospheric samplers, and atmospheric probes, all under the control of a base located on the planet-facing side of one of its major moons.

Figure 10-5. Spacecraft landing on Titan. An artist's concept of a Viking-type scientific package landing on Titan, a satellite of Saturn, which is seen in the background. Titan has a heavy atmosphere, hence the parachutes. (Courtesy of NASA)

Investigations of the moons would be directed from the base and, just as for the Jupiter system, would consist of mapping and geophysical surveys by artificial satellites supplemented by soil samplers and rovers and unmanned scientific data-collecting stations. Manned visits would be for specific surveys and investigations. The base's moon would be thoroughly surveyed by unmanned satellites, rovers, and samplers and by manned expeditions.

The effective temperatures of the satellite systems of the Jovian planets decrease with increasing distance from the Sun, going from 70K for Saturn to 60K for Uranus and 45K for Neptune. If the bases become large, they would have to be domed like the Callisto base. Since Jupiter and Uranus have been discovered to have rings, but far less grand than the rings of Saturn, it is possible that Neptune, too, has a ring. The ring systems would be investigated by unmanned satellites, which would bring back samples to the base as well as make measurements.

The candidate moons for base sites—Titan for Saturn, Titania or

Oberon for Uranus, and Triton for Neptune—present slightly different problems. Titan's diameter of 5800 kilometers (3600 miles) makes it probably the second largest planetary satellite in the solar system. Through a telescope, one sees a layer of reddish-brown clouds that cover the surface either partly or totally, visual evidence for an atmosphere. The atmosphere of Titan has a surface pressure of from 0.1 to 1 Earth atmosphere, and its surface temperature is 175K ($-144°F$), which is warmer than expected. This suggests that a greenhouse effect operates. The atmosphere of Titan is mostly methane and hydrogen in about equal parts, somewhat like that of the primitive Earth. Titan is therefore a world of great biochemical interest. Its surface gravity is 11 percent of Earth's while its average density if 1.35 gm/cm^3, implying that it contains much ice or hydrogen or both, with little rocky matter. Probably, Titan has a rocky core surrounded by a wet, rocky mantle. The crust may be water ice with much methane mixed in. If Titan's atmosphere is deep, a layer of liquid methane may be floating on the ice.

Since the seeing from Titan is obscured, it cannot be the site for a manned observatory for observing Saturn. A manned space station would be needed. Also, Titan may not have a solid surface, and there may well be major engineering problems in constructing and operating a base in its toxic atmosphere. It might be better to establish a base on one of Saturn's moons that has no atmosphere. The next most massive of the moons are Rhea, the fifth planetary satellite and the next in from Titan, and Iapetus, the eighth out.

Rhea and Iapetus have surface gravities only 2.5 percent of Earth's, so interplanetary spaceships could easily land on them. Iapetus is unusual even for the moons of the outer solar system, for its leading hemisphere reflects only about 4 percent of the sunlight that reaches it, making it darker than a blackboard. Its trailing hemisphere, in contrast, is as bright as snow. This is probably caused by a rain of particles from Phoebe, the outermost moon of Saturn, which is in a retrograde orbit. Both Rhea and Iapetus consist of ices mixed with rocky material. Rhea is 1450 kilometers (900 miles) in diameter and orbits Saturn at 527,000 kilometers (328,000 miles), while Iapetus is 1800 kilometers (1120 miles) in diameter and revolves around Saturn at 3,560,000 kilometers (2,200,000 miles).

Titania, the largest moon of Uranus, is the fourth from the planet. It has a diameter of 900 kilometers (560 miles), a density about 1.5 gm/cm^3, and a surface gravity about 3.5 percent of the Earth's. Oberon, the second largest, is the outermost of the moons of Uranus.

It is slightly less massive, smaller, and has a little lower surface gravity. Not much is known about Titania or Oberon. They probably rotate once for each revolution, lack atmospheres, and are at least partly icy.

Neptune's largest moon, Triton, is probably the largest and most massive moon in the solar system; its diameter is 6000 kilometers (3700 miles) and its mass is 2.8 times that of our Moon. It moves in a retrograde orbit about Neptune at the distance of 350,000 kilometers (220,000 miles). While its surface gravity is about that of our Moon, its density is less, 1.9 gm/cm^3, which implies that it contains a fair amount of ices or hydrogen compounds as well as rock. Since its surface gravity is comparatively high, like our Moon it probably will have it own orbiting space station to serve as a transfer point for cargo and passengers. Since Triton is closer to Neptune than Callisto is to Jupiter, or Triton to Saturn, or Titania to Uranus, it will be a superior observation platform.

Pluto is a very strange planet. Its surface temperature is only 46K (−350°F), so what could have been its atmosphere has been frozen, and its surface probably has lakes of methane ice. Pluto is smaller even than our Moon, since its diameter is only 3000 kilometers (1800 miles). Its mass is only 0.23 percent and its surface gravity only 3 percent of Earth's. James W. Christy of the U.S. Naval Observatory recently discovered that Pluto has a moon, which he named Charon. Charon orbits Pluto at a distance of 20,000 kilometers (12,000 miles). It is about 1500 kilometers (900 miles) in diameter, so it is about half the size of Pluto. Both Pluto and Charon rotate in the same period, 6.3867 days. This is the first case found in the solar system of synchronous rotation of a planet and its satellite.

With its weak gravity, Pluto would be an easy site for landing and launching spaceships. In its black sky the Sun shines as a very bright star, only 45 arc-seconds across. It is startling to realize that there is a body in the solar system from which the Sun appears starlike.

Some of the telescopes that will be located at various points in the solar system, whether in space or on the surface of airless moons, may be used part of the time to search for, and observe, nearby stellar systems. Our knowledge of the solar system will be the foundation for gaining understanding of planetary systems about other stars that will surely be detected in future years. The exploration and colonization of the solar system will be the prelude to the exploration and colonization of nearby planetary systems.

Notes

1. Carl Sagan, "The Planet Venus," *Science*, 133 (March 24, 1961), 849–858.
2. Jeffrey M. Lenerowitz, "Jovian Moon Volcanoes Detected," *Aviation Week and Space Technology*, March 19, 1979, pp. 14–20.
3. Morton W. Miller, Gary E. Kaufman, and H. David Maillie, "Pioneer 10 Jovian Encounter: Radiation Dose and Implications for Biological Lethality," *Science*, 187 (February 28, 1975), 738–739.
4. Krafft A. Ehricke, "A Long-Range Perspective on Some Fundamental Aspects of Interstellar Evolution," *Journal of the British Interplanetary Society*, 28 (1975), 722.

For suggested reading, see "The Solar System," *Scientific American*, September 1975, special issue.

11
SURVEY OF OUR GALAXY

When we look up at the sky on a clear, moonless night, far from city lights, we can see a circular band of light, the Milky Way. The Milky Way goes all around the celestial sphere, dividing it into two equal parts. It is broadest and brightest in Sagittarius, becoming dimmer and narrower on both sides. Also, the brightness decreases rapidly as we look away from it. With binoculars and telescopes we can see myriads of faint stars and dark nebulae superimposed upon its generally smooth background. This is our view of our own Galaxy, the Milky Way, from inside it.

Since our solar system is in a dust-enshrouded region of space, it has been difficult to determine the structure of the Galaxy. The dust in the plane of our Galaxy lets us see only one-tenth of the way toward the Galactic Center. However, studies of hot young supergiant stars and stellar associations, radio studies of the 21-cm line of hydrogen, and studies of other galaxies have shown that our Galaxy is most probably a spiral galaxy of Hubble type Sb or Sbc, similar to the Great

Spiral in Andromeda. Two spiral arms extend from opposite sides of the nucleus, which makes up a small part of the Galaxy. The arms spiral outward and divide into many side arms. Most of the galactic matter is within the arms, although there definitely is interarm matter. From the side, the ratio of arm thickness to galactic radius is less than that of a long-playing 33-rpm record.

The Galaxy has a radius of about 50,000 light-years. The Sun is within an arm about 28,000 light-years from the Galactic Center. At the Sun's distance from the Galactic Center, the separation between arms is about 6000 light-years. The arms are about 650 light-years thick and 1600 light-years wide. The Sun is 650 light-years from the central axis of the Local or Orion Arm, or 150 light-years from its inner edge. In our vicinity, the pitch angle of the arms, the angle between a tangent to a spiral arm and the perpendicular to the direction of the Galactic Center, is roughly 20 degrees. The Galaxy has a mass of about 2×10^{11} solar masses and has somewhat more than 2×10^{11} stars.

Since stars make up most of the mass of our Galaxy, their distribution in space largely determines its structure, with the gas and dust of interstellar space composing the rest. The core of the Galaxy is 10 light-years in radius. It is surrounded by a flattened, spheroidal nuclear bulge with a radius of 6500 light-years. The bulge does not have spiral arms and contains old stars. The disk with its spiral arms extends for 43,000 light-years from the bulge. The disk has the youngest stars in the Galaxy, including the Sun, together with dust and gas clouds, while the older stars form a halo around the disk. The halo is a great spheroidal region whose diameter is at least as great as the disk's.

Vast star clusters of tens or hundreds of thousands of stars are scattered throughout the halo. The masses of these globular clusters are so great that they are spheroidal. In the disk, another type of star cluster, the galactic cluster, typically contains from 20 to 2000 stars. The Pleiades are an example of a galactic cluster.

The central parts of the Galaxy rotate like a solid body, completing one revolution about the nucleus in the same time. Farther out, the stars do not behave in this manner. Those that are closer in take less time to make one revolution than those farther out. If this has been happening since the Galaxy was formed, why have the spiral arms not wound up by now?

The spiral arm pattern is caused by a compression wave that moves through the gas and the stars. Since the compression wave rotates more slowly than the stars around the Galactic Center, the density of the matter increases as it passes through. The density of the gas and dust increases enough to make the formation of stars possible. Thus, the form of the spiral arms is outlined by bright young stars. The density of matter in the spiral arms is slightly greater than in the interarm regions. Hence the gravitational force helps to maintain the spiral arm pattern, since the stars and gas remain there longer than would be expected on the basis of Kepler's laws.

In the initial stages of space exploration, humans will stay pretty close to the Sun, astronomically speaking. Later, they will want to travel from one stellar system to another. Since the density of stars decreases as we move away from the plane of the Milky Way Galaxy, the main lines of exploration will be in the Milky Way plane. We can assume that the conditions in interstellar space will be similar to those in the outermost reaches of the solar system, with suitable correction factors. The first direct measurements begin as the Pioneer spacecraft leave the region of space dominated by the solar wind, the Sun's domain.

Pluto is a cold and lonely place by Earthly standards, but it is a very nice, warm place by interstellar standards of darkness and near absolute zero. Pluto is some 40 times farther away from the Sun than we are, but the nearest star, Alpha Centauri, is around 6800 times farther away. Between Pluto and the Alpha Centauri system there probably are few interesting stopping-off places.

Interstellar space in our own spiral arm of the Milky Way Galaxy is an almost perfect vacuum with only one atom per cubic centimeter. In the interarm regions and in the galactic halo, the density is even less, possibly a third to a twentieth as much.

The density of interstellar matter near stars tends to be greater than far from them. The interstellar medium is in two different forms, clouds or nebulae and the intercloud regions. The density in the intercloud regions is about 0.1 hydrogen atom per cubic centimeter, while in the clouds it may be 1000 to 10,000 or even more per cubic centimeter. There are two main types of matter in the interstellar medium, gas and dust, which are well mixed except near hot stars. Dust cannot exist near hot stars, since it would be vaporized.

In the Local Arm, there are 25 to 50 grains of dust per cubic kilometer. Each grain's diameter is less than a thousandth of a millimeter. The dust density is correlated with the gas density. There are various kinds of obvious nebulae—dark, reflection, and emission—but most nebulae are nonobvious. Their dust dims starlight with distance, causing it to appear redder by preferentially absorbing the shorter wavelengths or blue light, and also polarizes the starlight.

It is incredibly cold in interstellar space, even in the hot intercloud regions, since the density of interstellar matter is extremely low. If a starship were not moving and not internally heated, its skin would chill to a few degrees above absolute zero. Because of the blackbody radiation that fills space (the relic of the creation of the universe some 13 billion years ago), it can never get colder than 3K. Starlight and the cosmic rays would also warm the starship a little.

We know nothing directly about the occurrence in interstellar space of bodies larger than dust grains and smaller than stars. However, it is possible to estimate the amount of mass per unit area in the plane of the Galaxy for the solar vicinity. After the stars and the interstellar matter have been accounted for, there is nothing left over. More important, no comet or meteor has been found whose orbit indicates that it comes from outside the solar system. Therefore, the density of solid bodies ranging in size from dust grains to planets is likely to be far less in interstellar space than in the inner solar system. Further, the vast majority of such objects probably are only slightly larger than the dust grains responsible for the interstellar extinction of light.

Stars are the major constituents of most galaxies, including our own. They exist in a wide variety of masses, temperatures, radii, and chemical composition. Our primary concern is with the stars that belong to the main sequence—that is, stars that are burning hydrogen in their cores into helium. This stage in the evolution of the stars accounts for about 90 percent of their life history from stellar birth to the formation of the death remnant, be it white dwarf, neutron star, or black hole. The stars slowly expand during their stay on the main sequence, and, immediately after, become larger and much more luminous. If life is to evolve, then, it must do so during the star's main sequence.

Table 11-1 lists some of the important properties of main-sequence stars of the solar neighborhood. Dr. Stephen Dole has found that stars with suitable life zones around them are the early F through early K stars that have masses from 1.4 times to 0.7 times that of the Sun. Our principal concern in stellar exploration will be with the F, G, and early K-type stars. The number of stars of a given type per unit volume of space increases with decreasing stellar mass. The Sun is among the most massive stars that can have life-supporting planets.

If a planet of a single star has the proper orbital characteristics—

namely, a reasonably circular orbit and radius that is suitable in terms of the star's luminosity—then the planet will remain in the star's life-support zone for its entire orbit. But two-thirds or more of all the stars in the Galaxy may belong to binary- or multiple-star systems. The criteria for stability of orbits require that in a triple-star system, two of the stars be relatively close while the third star orbits about the close pair. We can examine the possibility of life-supporting planets in a multiple-star system by considering whether the star is essentially isolated or belongs to a close binary pair. If the binary stars are well separated, then the situation is similar to that of a single star. If the

TABLE 11-1. Types of Stars

Spectral Type	Color	Temperature (°K)	$M/M_\odot$	$R/R_\odot$	$L/L_\odot$
O5	Bluish white	40,000	40	18	500,000
B0	Bluish white	28,000	18	7.4	20,000
B5	Bluish white	15,500	6.5	3.8	800
A0	White	9,900	3.3	2.5	80
A5	White	8,500	2.1	1.7	20
F0	Yellowish white	7,400	1.7	1.4	6.3
F5	Yellowish white	6,600	1.3	1.2	2.5
G0	Yellow	6,050	1.1	1.05	1.3
G5	Yellow	5,500	0.93	0.93	0.79
K0	Orange	4,900	0.78	0.85	0.40
K5	Orange	4,150	0.69	0.74	0.16
M0	Red	3,500	0.47	0.63	0.063
M5	Red	2,800	0.21	0.32	0.008
M8	Red	2,400	0.10	0.13	0.0005

Spectral Type	Lifetime (Years) on Main Sequence	Life Zones (AU) Inner Radius	Life Zones (AU) Outer Radius
O5	1.6×10^5		
B0	1.7×10^6		
B5	3.6×10^7		
A0	2.8×10^8		
A5	1.1×10^9		
F0	2.0×10^9	1.82	2.62
F5	4.6×10^9	1.24	1.78
G0	7.5×10^9	0.88	1.26
G5	1.2×10^{10}	0.64	0.96
K0	2.1×10^{10}	0.60	0.66
K5	3.0×10^{10}		
M0	9.6×10^{10}		
M5	1.1×10^{12}		
M8	1.0×10^{13}		

Note: The Sun is a G_2 star. The table shows, besides spectral type, color, and the temperatures of the various classes of stars, their masses, radii, and luminosity relative to the Sun. It also shows the life zones in terms of astronomical units.

The data in this table is reprinted with permission from S.H. Dole, Habitable Planets for Man (New York: Blaisdell Publishing Co., 1964), and C.W. Allen, Astrophysical Quantities, 3rd. ed. (London: Athlone Press, 1973).

stars are close together, then planets may be able to have stable orbits about both stars in the life support zone. In the intermediate case of stars separated by a half to a few astronomical units, no stable orbits may exist in the life support zone.

The Milky Way Galaxy, with its billions of stars, is only one of a small group of about twenty galaxies, known as the Local Group. The largest is the Messier 31, the Great Galaxy in Andromeda, which is about 2 million light-years away. It is about twice as massive as our Milky Way Galaxy, which is the second largest member of the group. The other galaxies are much smaller. The Local Group fills an ellipsoidal volume of space about 2.9 million light-years in length and about 2.6 million light-years along its other two axes.

There are similar groups of galaxies at a distance about four times the size of the Local Group. The largest group of nearby galaxies is the Virgo Cluster, about 36 million light-years away. The Local Group and nearby clusters appear to be outlying members of a supercluster of galaxies centered on the Virgo Cluster. These superclusters are the fundamental organizational units of galaxy groups in the universe.

12
PROPULSION: THE KEY TO INTERSTELLAR FLIGHT

Can we reach the stars? This question has been debated and analyzed by a good many scientists and engineers since at least the early 1950s, when rockets began to be employed for scientific investigations, starting with the upper atmosphere. Interstellar flight is extraordinarily difficult to achieve in view of the immense distances, the almost incredible velocity demanded of the starship and the unprecedented requirements for energy. Some scientists believed that it was impossible, at least for the foreseeable future. Others, more optimistic, came upon highly perplexing questions.

Feasibility of Interstellar Flight

We will first consider the views of some of the critics of interstellar flight, then those of several of its proponents, and try to see if there is a way that may lead to its achievement—assuming, of course, that science and technology will advance to the level that will make it feasible.

Serious consideration of interstellar flight began before the onset of solar system exploration. Dr. Edward Purcell[1] of Harvard University tried to prove the impossibility of space flight beyond the solar system in a lecture he gave at the Brookhaven Laboratory in 1961. Assuming, he said, that an ideal nuclear fusion propellant was used—that is, hydrogen completely fused into helium—an exhaust velocity of one-eighth that of light could be reached. But if the maximum speed of the spaceship were to be 99 percent that of light, the mass ratio would have to be 1,600,000,000 to 1 with 1-g acceleration and deceleration and return.

Only matter-antimatter annihilation could do better than fusion. The exhaust velocity would be nearly that of light and the mass ratio would be only 14 to 1. However, this would be only for a one-way flight without stopping or deceleration. If the spaceship were to slow down and stop at its destination, return, and decelerate to a halt in our solar system, its mass ratio would have to be 40,000 to 1. According to Purcell, then, interstellar flight would appear impossible.

In an article in *Science*, "The General Limits of Space Travel," Dr. Sebastian von Hoerner[2] of the National Radio Astronomy Observatory strongly attacked the idea of interstellar flight and made a detailed case against it. Like Purcell, von Hoerner maintained that for interstellar travel, the ship's velocity must be near that of light. Since only the annihilation of matter could supply the enormous energy needed, and since matter could not be wasted for use as thrust, only photon thrust, the thrust of light, could be employed.

Given that assumption, von Hoerner found that the ratio of the power of the engine to the total mass of the starship would have to be 4000 horsepower per gram.

Von Hoerner went on:

Or to express it another way, . . . the engine of a good car, producing 200 horsepower, could not weigh more than 50 milligrams—one tenth of the weight of a paper clip."

Gigantic obstacle piles upon gigantic obstacle. Power must be converted into light to serve as propulsion.

A large transmitting station of 100 kilowatts power output can then give the tiny thrust of 30 milligrams and so can an aggregate of searchlights

> with combined power of 100 kilowatts. And all this should not weigh more then 1/15 the weight of a paper clip. The power source and transmitter requirements must be combined and the mass entering . . . must contain reactor as well as emitting stations.

This is depressing enough, but worse follows. Von Hoerner pursued his thesis relentlessly. As an example, he assumed a spaceship with a payload of only 10,000 kilograms (22,000 pounds) with another 10,000 kilograms (22,000 pounds) for power plant plus emitters (of light for thrust). For a velocity of 98 percent of light, the ship needs a mass ratio of 10:1 and its total mass would be 200,000 kilograms (440,000 pounds). To accelerate at 1 g, it would have to have an engine with the power of 600 million megawatts. For comparison, present nuclear power plants generate about 1000 megawatts. Von Hoerner summed up:

> We would need 40 million annihilation power plants of 15 megawatts each, plus 6 billion transmitting stations of 100 kilowatts each, altogether having no more mass than 10 tons, in order to approach the velocity of light to within 2 percent within 2.3 years of the crew's time.

He concluded:

> The various questions dealt with in this article have not led to the definitive answer that interstellar flight is absolutely impossible. We have found simply the minimum travel times given by different assumptions, and we have found the requirements for reaching these limits. This is, at present, all we can do, and the final conclusion is up to the reader. The requirements, however, have turned out to be such extreme ones that I, personally, draw this conclusion: space travel, even in the most distant future, will be completely confined to our own planetary system, and a similar conclusion will hold for any other civilization, no matter how advanced it may be. The only means of communication between different civilizations thus seems to be electromagnetic signals.

Dr. Ernst J. Öpik[3] of Armagh Observatory took up the controversy in a paper published in 1964. He vigorously criticized both the concept of interstellar flight in general and Bussard's proposal, made four years before, that the interstellar gas could be used as fuel for the ship's fusion engine. The starship, Öpik maintained, would have to have an intake of 50 kilometers (31 miles) in radius, even in the most favorable regions such as the dense interior of the Orion Nebula, for a payload of 1000 tons. The intake, moreover, would need to be a cobweb of steel wires with a mass of 16,000 tons that would collapse under its own weight. If more than 0.001 percent of the kinetic energy of the interstellar gas were lost on contact with the intake, the speed of the ship would be reduced, not increased, by the inflow of the gas. For the ramjet to function effectively, the density of the interstellar gas would have to be so high that the wires would be melted by heat long before the required density would be reached. Öpik decided that interstellar flight is impossible, at least at 98 percent of the velocity of light.

The Project Cyclops report, which was devoted mainly to a proposed radio telescope for searching for extraterrestrial intelligence, gave some attention to other possibilities for contacting intelligent life outside the solar system, including interstellar travel. As a test for achieving the latter, Dr. Bernard Oliver[4] postulated a rocket ship with an annihilation engine that would convert all the matter to light. The rocket ship would almost reach light velocity and take only 10 years to travel to Alpha Centauri, 4.3 light-years distant, explore its solar system (assuming it has one), and return. The ship would accelerate at the start of its journey, decelerate before reaching Alpha Centauri, and do the same on its return voyage. Oliver calculated that the spaceship would weigh 34,000 tons at the start, carrying 33,000 tons of matter-antimatter fuel, which at only 0.1 cent per kilowatt-hour would cost $1,000,000 billion. The ship's engine would have to have a power of about a trillion megawatts to accelerate the ship to cruising speed. If all of this power were converted into thrust, except for one part in a million, which would be astonishingly efficient, the ship would somehow have to absorb a million megawatts of heat. Oliver remarked that a million megawatts of cooling in space would require about 1000 square miles of surface if the radiating surface were to be at room temperature. He concluded:

> *A sober appraisal of all the methods so far proposed forces one to the conclusion that interstellar flight is out of the question*, not only for the present but for an indefinitely long time in the future. It is not a physical impossibility but it is an economic impossibility at the present time. Some unforeseeable breakthroughs must occur before man can physically travel to the stars.

Are the critics of interstellar flight correct? Should we agree with Purcell's verdict on interstellar flight: "All this stuff about traveling around the universe in space suits—except for local exploration, which I have not discussed—belongs back where it came from, on the cereal box."[1]

The case against interstellar flight is strong, as we have seen. To cross the vast distances to even the nearest star, incredible power will be needed. Alpha Centauri is 260,000 times as far as the Sun. Our solar system is only a speck in the space ocean.

Pioneer 10, the first spacecraft to reach Jupiter, took 21 months for its flight of 998 million kilometers (620 million miles), averaging 18.2 kilometers (11.4 miles) a second. Carl Sagan, who composed the message engraved on one of its panels, calculated that after leaving the solar system at a speed of 11.4 kilometers (7.1 miles) a second, slowed down by the Sun's gravitational field, it will take 80,000 years to travel one parsec, 3.3 light-years. Thus, if Pioneer 10 were sent to Alpha Centauri, it would take 104,000 years to get there.

A light-year measures both time and space, since light travels at constant speed in a vacuum. It is the distance covered by light traveling at 299,695 kilometers (186,262 miles) a second for one year—9,460,488,000,000 kilometers (5,877,863,000,000 miles). According to the special theory of relativity, the speed of light is the celestial speed limit. It cannot be exceeded, for as the speed of a body increases, so does its mass. This effect becomes apparent as the body's speed reaches that of a substantial fraction of light velocity. A spaceship, therefore, can never reach the speed of light—for to do so, its energy would have to be infinite. Despite this, starships must attain a sizeable fraction of light velocity to reach the nearer stars in a human lifetime. This is doubly necessary if they are to return to Earth.

Can we attain the level of technology that will enable mankind to voyage far out into the space ocean, or must we remain forever confined to our solar system Mediterranean? Arthur C. Clarke, who is both a physicist and a science writer, strongly disagreed with the Purcell-Öpik-von Hoerner school. In the January 16, 1968, issue of *Science* he wrote: "Any really competent extrapolation shows interstellar travel to be a rather simple engineering accomplishment to be expected within a mere two or three centuries of control of thermonuclear fusion."[5]

Who is right? Interstellar flight is certainly not a simple engineering accomplishment! The question is whether technological advances can increase velocity from our present 16 kilometers (10 miles) a second with chemical engines to, say, 90,000 kilometers (56,000 miles) a second, 30 percent of the velocity of light. Fusion is making fast progress, and a commercial fusion reactor could be operating in the 1990s. The transition from a laser fusion reactor to a laser fusion rocket engine may be smooth and rapid with a reasonably supported effort. The first fusion-powered spaceship could even be in service by the end of this century. Assuredly, this issue will not take two or three centuries to resolve. It should be settled by two or three decades of intensive research and development, quite possibly by 2025. If Nature's answer is "No," the human race will remain confined to the solar system. If her answer is "Yes," then the universe will be open to humankind.

Von Hoerner's objections to the possibility of interstellar flight are well founded. We readily agree with many of them, such as chemical fuels being out of the question. They are much too feeble for more than the preliminary exploration of the solar system. (See Chapter 5.) Fission of uranium-235 could generate an exhaust velocity a twenty-third that of light, 13,000 kilometers (8060 miles) a second, while fusion could push it to 59,800 kilometers (37,252 miles) a second, one-fifth of the speed of light. These figures are for a spaceship with a mass ratio of ten to one.

However, initially, von Hoerner assumed the most extreme case by arguing that only a speed close to that of light would do for interstellar flight. Since energy requirements soar steeply as speed rises from 30 to 90 percent of light velocity, he made the problem much worse than it need be.

Later on, von Hoerner changed his mind about the possibility of interstellar flight. The Orion project on a starship driven at one-third of light velocity by exploding hydrogen bombs opened the way, he believed, to interstellar flight, even with present technology. He was also persuaded by the idea of a space ark that would make interstellar travel possible, although it would take generations to traverse interstellar distances.

Maxwell Hunter, an aerospace engineer, explained this energy requirement clearly.[6] Suppose, he said, that an interstellar spaceship with a mass of 4500 kilograms (10,000 pounds) accelerates at 1 g until it reaches 30 percent of light velocity. To do so, its engine would have to generate 1356 megawatts, equivalent to 4.4 billion newtons (one billion pounds of thrust). Huge, indeed, but only 83 times the thrust of a Saturn V rocket on takeoff and not much more than the average power of an American nuclear powerplant. To reach 97 percent of light velocity, however, the ship's engine would have to generate 10^{18} times as much energy or 10^{21} megawatts, which is equal to the power radiated by the Sun.

How can we cross the interstellar chasm? Daunted by the sheer distance, some have proposed a space ark, an artificial miniature Earth traveling so slowly that it would take centuries to reach the nearest star. For example, Dr. Gregory L. Matloff[7] of New York University has suggested modifying one of O'Neill's space colonies to serve as a space ark. The ark would be two gigantic cylinders, each 1000 meters (3300 feet) long and 100 meters (330 feet) in diameter, attached together, with the engine and fuel tanks in between. The pulsed fusion engine would have a mass of 500,000 kilograms and the fuel tanks would hold 150 million kilograms of deuterium and helium-3. Suppose the space ark leaves the solar system traveling at 1 percent of light velocity, 2990 kilometers (1860 miles) a second, on January 1, 2100. It arrives at Alpha Centauri on January 1, 2530. In the journey of 430 years, more than 13 generations have been born and passed away. Even Methusaleh would be daunted by the prospects of traveling in a space ark.

The space ark advocates fail to see that they have exchanged space for time. They have made interstellar travel more difficult, not easier. People have been content to live on Earth because it is so vast compared to a human being and seems endlessly varied. Until the twentieth century, men could not conceive of living anywhere else. Human beings would be happy living in a space ark only if their lives would appear as free and satisfying to them as life on Earth. Existence in an interstellar space ark would, we believe, be imprisonment for them and their descendants. Such an existence would prove intolerable.

To alleviate the time travel problem, Matloff later suggested that the space ark could be converted into a ram-augmented interstellar rocket that would have its fuel supply supplemented by gases collected from interstellar space. Why, then, construct a space ark at all? Why not go all-out and try to realize the true starship that could travel to the stars in years, not centuries?

Science and technology have reached the level at which they can often offer a variety of solutions to major technical problems. In the space enterprise, the psychological factor is important in choosing an approach. The interstellar space ark is feasible, but who would want to live in it? We do not believe that an interstellar space ark will ever be launched. Progress is bound to overtake it. Suppose that in the year 2025 a space ark takes off from lunar orbit bound for Alpha Centauri. Twenty-five years later, as it is sluggishly moving along, a starship designed with the advantage of 25 more years of technical advances reaches it and docks. The starship's captain invites the ark's crew and passengers to transfer to the starship, which is scheduled to reach Alpha Centauri in five years, not in 375 years. Who would remain in the space ark?

Hibernation and Interstellar Flight

Many writers have argued that the best way to avoid the dilemma of generating enormous power for the starship or enduring the endless voyaging of the space is hibernation. Let medical scientists invent artificial hibernation for humans and lo!, the problem is solved. Once the interstellar argosy is on its course, the crew hibernates, automatic machinery takes over responsibility for the ship, and, almost before they know it, they are awakened and their destination is only a short distance away.

The idea is enticing, but it will not work. Humans are not natural hibernators.[8] There is not a single instance in all history of any human being hibernating. Hypothermia, lowering the body temperature of a patient for surgery, is *not* hibernation. When a mammal such as a ground squirrel hibernates, the animal unconsciously disengages the controls by which its body temperature is kept constant and lets it drop to nearly that of the surrounding air. The heart rate and metabolism similarly decrease. Somehow, the ground squirrel's systems prepare it for hibernation. Also, the animal can voluntarily awaken and come out of hibernation.

Contrast medical hypothermia with this natural process. Medical hypothermia is a closely controlled procedure used mainly for heart operations to lower blood pressure, control hemorrhage, and so reduce the damage to tissues, especially the brain, that are sensitive to reduction of oxygen supply. It is a drastic, dangerous process employed only when necessary. The patient is continually monitored by sophisticated instruments and a team of surgeons, nurses, and medical technicians. If the patient's heart starts to fibrillate or beat irregularly, his temperature is immediately raised by diathermy coils around his abdomen.

There is no drug known that can induce human hibernation. Moreover, if such a drug were discovered, its effects, administered over a period of years, not minutes or hours, might well wreck the subject's nervous system. Consider what a comparatively mild psychotoxic drug such as alcohol does to the brain.

What advantages are propounded for hibernation? Supposedly,

they are as follows: reducing the mass of the ship's life support system, retarding aging, protecting the crew against harmful radiation and diseases, and, best of all, reducing the psychological stress of space existence, which would result in boredom, depression, poor sleep, and so on. None of these contentions has any validity.

Any artificial hibernation system would have to be multiply redundant, for the lives of the crew would depend upon it. Any mass saved by reducing the ship's life support system would be more than offset by the added mass of the hibernation system.

Aging is a medical field that only now is getting serious attention by the medical scientists. By the time that the first starship is ready for its voyage, human biology should have means for reducing its ill effects. What the medical scientists on board would be concerned with would not be the appearance of the crew but maintaining them in top physical and mental condition over the years.

The claim that existence in space induces stress is due to misconceptions about the nature of space travel. Boredom is internal, not external. None of the astronauts or cosmonauts have complained of being bored while on a space mission. On the contrary, the astronauts often complained to their mission controllers that they had too much to do and not enough time to relax. We believe psychological stress will not be a hazard of interstellar travel.

Artificial hibernation, then, would offer no advantages and would be so hazardous to interstellar flight that we believe it will not be considered.

Project Orion

The first serious starship study was Project Orion, which began as a spaceship for solar system exploration. Orion, a giant spaceship, was to be built with then-existing technology. It would be propelled by atom bombs. Conceived in 1955 by Stanley Ulam, it received little attention until 1958, after the Soviet Union had beaten the United States to opening the Space Age by launching its first Sputnik.

Theodore T. Taylor, the design team leader of Project Orion, was a physicist who had invented several new types of atomic bombs. He was worried by the belated American attempts to catch up with the Russians by merely sending a few kilograms of satellite into orbit. He believed that the United States should start exploring space on a grand scale by launching spaceships worthy of the name. Orion was to be bullet-shaped, 41 meters (135 feet) in diameter and over three times as long. It would be driven by ejecting and exploding fission bombs, each with a force of 20,000 tons, that of the atom bomb that exploded over Hiroshima.

The next year, 1958, the Advanced Research Project Agency funded a preliminary study of this startling idea. Orion was to be fueled with 2000 atom bombs that would be expelled and then exploded about 30 meters (100 feet) behind the spaceship, whose bottom would be protected by a heavy "pusher" plate to take the shock of the successive blasts. (See Figure 12-1, page 204.) What would prevent the crew from being shaken to bits? If the reader has ever stood near a heavy shock machine and felt the abrupt shock wave surge upward through his body as the hammer slams into the anvil, he will know that this question is far from facetious. The solution proposed was a set of 15-meter (50-foot) shock absorbers connecting the pusher plate to the ship's bottom. The crew would also be protected by 60,000 kilograms (65 tons) of shielding.

If the first Orion worked, why not follow it with an advanced version? Taylor sketched a Super-Orion 1.6 kilometers (1 mile) in diameter to be propelled by a million hydrogen bombs on its voyage to a nearby star. Then, in 1963, the United States signed a treaty with the Soviet Union that bound the two nations to stop conducting tests of atom bombs in the atmosphere and in space. This put an end to Project Orion. The story of its demise is not that simple, however, and it is instructive—for it sheds light on how a scientific project can be ended for nonscientific reasons.

Dr. Freeman J. Dyson of the Institute for Advanced Study had joined the Project Orion team because he strongly believed in its scientific utility. He ascribed its ending to a singular combination of events.[9] The Advanced Research Projects Agency had been set up by the Department of Defense after World War II to support science programs that had military aspects. The Orion Project had been started by the General Atomics Company in 1958 and was being run by nuclear weapons experts. NASA had not yet been established and ARPA was the only Government agency that could be approached. ARPA granted a contract for preliminary studies and, while it was a military agency, did not insist that Orion should be called a military project. Shortly afterward, NASA was established. NASA moved quickly to take over all of ARPA's civilian space projects, for it had been given responsibility by Congress for all nonmilitary space activities. Orion was soon ARPA's only space project.

In 1960, ARPA decided to drop Orion. Taylor asked the Air

Force to back it. Logically, there was and is no military need to explore the planets, but the Air Force agreed. This left Project Orion in a very awkward position. Seemingly military, it was actually civilian, while the Air Force had to justify the Project Orion budget in military terms. After a while, the Secretary of Defense told the Air Force to ask NASA to take it over. NASA became interested in 1963, but by then it was too late.

In effect, Orion had become a science orphan that no one wanted to adopt. A private firm would not have the funds or the prospect of return to sustain the enterprise. The Air Force could not logically support a large civilian space project that was now within NASA's jurisdiction. NASA did not want it, since the public was anti-bomb. The backers of the atmosphere test ban treaty were opposed to it, mistakenly believing that it was a nuclear weapons project. Yet for all that, the work of Taylor, Dyson, Ulam, and their colleagures was not wasted—for scientific work is wasted only when it is kept secret. Project Orion led to a more practical concept for a starship.

Von Hoerner was persuaded by the Orion project that interstellar flight was possible, although extremely unlikely.[10] Dyson in a paper published in 1968[11] showed that a spaceship propelled by ejecting and exploding hydrogen bombs at a safe distance could reach a velocity of about one-thirtieth the speed of light, about 10,000 kilometers (6000 miles) a second. This would still be very slow for interstellar distances, but it showed that interstellar flight is possible.

Nuclear fusion was making slow progress in the early sixties. The approach was to try to confine a mixture of deuterium and tritium long enough, about one second, at a high enough temperature, about 100 million degrees Kelvin, and at a sufficient density, about that of the air in the stratosphere, so that enough deuterium and tritium nuclei would fuse to give off more energy than had been pumped in. The fusion researchers had much trouble in trying to keep the gas, a plasma, from escaping the magnetic field that confined it.

Since lasers were coming into use at this time, some physicists thought that they might be the key to fusion. To avoid the problems of magnetic confinement, why not have powerful lasers smash pellets of frozen deuterium-tritium? A decade of research on laser fusion has so advanced the state of the art that apparently the last major barrier to successful fusion in this manner will be breached by the development of sufficiently powerful lasers. To overcome the mutual repulsion of the tritium and deuterium nuclei so they will fuse, the laser beams must strike the pellet and compress it in a billionth of a second. Unlike the magnetic-confinement fusion reactor, the laser reactor does not have heavy electromagnetic coils, so it should be lighter and more suitable for a spaceship engine.

Laser-powered Starships

But why have an engine in the spaceship at all? If the power source could be kept on Earth and the power delivered to the space vehicle, the latter could be propelled without an engine or fuel. The possibility is tempting.

In 1975, NASA's Lewis Research Center began to investigate beamed-power spacecraft. Powerful lasers on the ground would generate beams that would be aimed at an orbiting spacecraft. After passing through the spacecraft's window, the beams would heat gas inside the machine, probably hydrogen seeded with particles to absorb the beams. The heated gas would shoot out of the spacecraft's rocket nozzle. All that the spacecraft need carry would be a gas tank. Its propulsive energy would come from a ground station that could be luxuriously big and heavy. In effect, the engineer would leave his problems on the ground.

Like everything else, however, the beamed-power spacecraft has its disadvantages. Lasers of the order of 100 kilowatts are required, and they must run continuously—unlike most lasers, which fire in bursts. The beam must be very narrow and concentrated. Attenuation by the atmosphere must be overcome, and the ground laser must be in sight of the spacecraft. For a specific impulse of 1000 seconds, the spacecraft rocket chamber must reach a temperature of 3000K. Despite the effects of the laser beam, the spacecraft window must remain transparent for the lifetime of the spacecraft, perhaps as long as 10 years. A diamond window might be best! The propellant gas must be opaque in order to absorb the laser beam, or it will not heat up.

The best location for a beamed-power station might be in space. A space station could have huge mirrors to transmit laser beams to satellites and spacecraft over distances of thousands of kilometers through the transparent near-vacuum of space. NASA may use beamed power to speed spacecraft to planetary trajectories.

If laser beams could impel a small spacecraft to Mars, why not construct a giant laser and propel an interstellar probe or even a starship to a nearby star? This would eliminate the fuel handicap. Dr.

Robert L. Forward of Hughes Research Laboratories, who first proposed using lasers to propel spacecraft in 1962, used the simple equation $S = R\lambda/D$ to show that the spot size S of a laser with diameter D can be as small as a light sail, even at long ranges R, provided the wavelength λ is short enough, and the transmitting laser diameter is large enough. For example, a ring of 100 blue lasers in a circle 1000 kilometers in diameter, kept phase-locked with a pilot beam from the light-sail ship, would produce a spot size of 20 kilometers at the 4.3-light-year distance of Alpha Centauri.[12]

The interstellar laser beam has been discussed by several scientists. Professor G. Marx[13] of Roland Eotvos University proposed propelling an interstellar vehicle with laser beams, but Professor J. L. Redding[14] of Bishop's University countered that the power specified, about a billion megawatts, would be so vast and the space probe so light that it would not work. Also, as the probe speeded up, the Doppler shift would become more pronounced; the light waves coming in would lengthen toward the infrared and become weaker. The scheme would be self-defeating. What may work for small spacecraft near the Earth or even in the solar system cannot necessarily be extrapolated to the vast reaches of interstellar space.

Since the exchange by Marx and Redding in the mid-1960s centered on the use of continuously transmitting lasers, more powerful lasers, especially for fusion research, have been developed. Conceivably, an extremely powerful arrangement of giant lasers could be constructed in space. One cannot be dogmatic about the future of space lasers—or of any field of technology, for that matter. However, the beam-propelled starship has a special drawback: it must remain in the beam to receive power, so it cannot cruise at will in space. Conversely, the beam must always be on target, and this imposes very strict tracking and pointing requirements.

What if light itself provided the thrust? Light exerts pressure. For example, sunlight sweeps tiny particles of dust out of the solar system. If the thrust measured up to its task and if the rocket's exhaust were a beam of light, the rocket could almost match light velocity. In the late fifties and early sixties, several physicists studied the feasibility of a pure photon or light-beam rocket. They concluded that the idea was interesting but impractical. Dr. Eric Stuhlinger of the Marshall Space Flight Center analyzed the issue incisively.[15] His conclusions still stand.

You can buy a photon-powered rocket for a dollar or so. It is called a flashlight. Turn it on. Does it push against your hand? It does, but the thrust of the light beam is too minute to be felt. If, Stuhlinger said, we took a flashlight out into space and turned it on, it would propel itself. By the time that its battery was exhausted, the flashlight would be speeding along at 0.01 centimeter (0.039 inch) per second, which is equal to 3.6 centimeters (1.4 inches) per hour.

The thrust of a rocket engine is its exhaust velocity per second times the mass expelled per second. Light is made up of wave packets called photons, which can provide thrust when they transfer momentum. The aerospace engineer faces an almost insoluble problem in designing a photon engine. How do you get thrust into the light beam? The thrust of a light beam is the total power in the beam divided by the speed of light, while the power is equal to the rate of conversion of mass (by nuclear reactions) into radiant energy times the speed of light.

Let us suppose that science learned how to convert mass totally into energy and a pure photon rocket were built. How would it perform? Stuhlinger gave an example. A starship is to be driven steadily by its photon engine for one year. It has a mass of 50,000 kilograms (110,000 pounds) minus its fuel. At the end of the year it will have reached nine-tenths light velocity. Since it had a total mass of 217,500 kilograms (478,500 pounds) at launch, 167,500 kilograms (368,500 pounds) of fuel will have been converted into photons in that period. This conversion of mass into energy would be equivalent to a total power output of 475 million megawatts. (In 1957 the total electrical power output of the world was 350,000 megawatts.) Also, the temperature would rise to 10 trillion degrees Kelvin if only half the mass of fuel were converted into energy. How could the ship be kept from vaporizing? This example suggests that speed close to the velocity of light will be extremely difficult to attain.

Gravity-assisted Spaceflight

Gravity machines might be used to speed up spaceships. Professor Dyson described how a double star could be used as a gravity machine. If one of a pair of stars were a small but massive white dwarf, a spaceship swinging around it would be accelerated to a speed of 2000 kilometers (1240 miles) a second by an acceleration 10,000 times that of Earth gravity without any damage to the spaceship or its passengers!

NASA has used this slingshot technique to send a Mariner spacecraft skimming around Venus and on to Mercury. The technique has also been used to swing a Voyager around Jupiter and on to Saturn, and doubtless it will be a standard maneuver in future multiplanet missions. The gain, however, is only a few kilometers a second—excellent for the solar system but not in the least helpful for interstellar flight.

Krafft Ehricke[16] has described a plan for an ultraplanetary probe, a spacecraft that would explore space beyond the solar system out to about 6000 astronomical units or a tenth of a light year. This could be the start of interstellar exploration. The vast reach between Pluto and the nearest stars is not a void without scientific interest. Here the comets spend their eons. They may be the best-preserved samples of the solar nebula from which the solar system was formed. The interstellar medium may have an important influence on life on Earth. When the Sun enters a dense cloud region, the volume of space that the Sun dominates via the solar wind may be compressed and the transparency of the inner solar system reduced.

Ehricke's plan is to have two planets and the Sun cooperate in hurling the probe clear out of the solar system. The spacecraft would be launched from Earth and head toward Saturn. It would swing around that huge planet and, behind it, fire its engine, which would send it into a retrograde solar orbit (a clockwise orbit as seen from above the north pole). Speeding inward, it would pass near Jupiter and be hurtled around it toward the Sun. Rushing around the Sun, only 1.6 million kilometers (a million miles) above its photosphere, the spacecraft would be swept out of the solar system at 600 kilometers (370 miles) a second. In 50 years it would be a tenth of a light-year or 6320 AU from the Sun. While 600 kilometers a second is almost 100 times as fast as today's spacecraft, it is crawling compared to the speed of light. For faster speeds, Ehricke said that the gravity-assist method is of little use. To achieve near-light velocity, a starship can resort only to energy. The only possibilities are fission, fusion, and the annihilation of matter.

Fusion-powered Starships

Fission can convert up to 0.12 percent of the fissioning mass into energy and reach a maximum specific impulse of 1.5 million seconds, while fusion can convert up to 0.9 percent of the reacting mass into energy with a maximum specific impulse of 4.1 million seconds. Fusion is, therefore, preferable for powering a starship. (Fission propulsion for solar system flight was discussed in Chapter 5.) One of the first studies of fusion for interstellar flight propulsion was made in 1966 by Dwain F. Spencer[17] of the Jet Propulsion Laboratory. Research on controlled nuclear fusion was at an early stage and success seemed decades away. Spencer sketched a fusion engine that would burn deuterium with helium-3. The mixed gases would be pumped into a cavity surrounded by superconducting magnetic coils, which would confine the extremely hot plasma so it would not touch the cavity wall and so end the fusion reaction. The ultrahot gases would flare out the open end of the cavity and through the nozzle to drive the starship.

To attain three-tenths of the speed of light, all the helium-3 and deuterium would have to fuse, which is no easy matter. Even for a one-way mission, the starship would have five stages, starting to decelerate when only halfway through its voyage. At takeoff, the ship would weigh ten thousand times as much as it would on arrival. If the final stage were to be only 100 tons, at launch the ship would have to weigh one million tons. And, for all that, the starship would crawl through space, accelerating at 0.004 g to reach a star 5 light-years away in 50 years. Spencer concluded that fusion would be marginal for interstellar flight but remarkably promising for solar system missions.

By 1975, lasers had been so greatly improved that they were being investigated for use in a new type of fusion reactor, the inertial confinement reactor that we have briefly discussed. R. Hyde, L. Wood, and J. Nuckolls of Lawrence Livermore Laboratory prepared a plan that would take advantage of laser fusion progress and be an advance beyond Orion.[18] They proposed using little frozen pellets of deuterium-tritium, weighing 0.015 grams each, instead of fission bombs. Five hundred pellets a second would be exploded by lasers. Two-thirds of the tiny fireball's energy would be in charged particles that would be collimated into the exhaust by magnetic coils. As the expanding fireball pressed on the magnetic field, current would be induced in a pickup coil. The induced current would flow into a capacitor, which would discharge and fire the lasers. In this way, the engine cycle would be self-sustaining.

The Hyde-Wood-Nuckolls engine would consume 631 kilograms (1425 pounds) of deuterium and tritium each day and have a thrust of 24.5 billion newtons (5.5 billion pounds). Its specific impulse would be 333,000 seconds, whereas that of today's chemical engines is about

450 seconds. The theoretical maximum impulse from deuterium-tritium fusion is 2,640,000 seconds, so that the H-W-N engine taps only 12.6 percent of its possible maximum energy. The gain of output energy over that used to raise the gas temperature to fusion level is about 100. Dr. T. A. Heppenheimer,[19] who was then at the Goethe Institute, calculated that the D-T specific impulse could be raised to 1,000,000 seconds.

What could be done with this engine? It could power a single-stage interstellar probe that could attain one-tenth of light velocity, 29,900 kilometers (18,600 miles) a second. But Heppenheimer believes that it could be best used by accelerating to 10 percent of light velocity and flying past the target star at that speed. This flyby mode is not necessarily the least rewarding. A probe could fly past one star and be diverted by its gravitational field to another not far off its course. For example, a probe could be sent to Barnard's star, 5.9 light-years distant, reach it in 64 years, and then be sped to 70 Ophiuchi, which it would reach 92 years later.

Project Daedalus

Enter the British Interplanetary Society. This society has stimulated analysis of the innumerable problems of space flight since its founding in 1933, when firing a rocket to 160 kilometers (100 miles) altitude seemed an incredible feat. The society's interest in interstellar flight began after World War II, and it published its first paper on interstellar exploration in 1952.

In 1972, Dr. Alan Bond suggested forming a working group of BIS members to design a practical starship. This had never been done before. The suggestion was enthusiastically received, and on January 10, 1973, a meeting of the society was held in London to discuss the project.[20] Bond pointed out that while numerous papers on interstellar flight had been published, they were narrowly specialized. An integrated study of an interstellar mission would be useful in determining its feasibility. He recommended choosing the simplest possible one, a stellar flyby, since even the easiest would be formidably difficult. If the probe were launched in the year 2000 on a flight to Barnard's star, the probe would reach it in 40 years. Some of the younger men and women in the audience of 120 would then be alive to see the results of their labors.

By July of the following year, six committees were hard at work. The physicists and engineers were all volunteers who worked on the project in their spare time. In accordance with the British fondness for the Greek classics, the project was named Daedalus (literally, "cunningly wrought").

The first decision was on the type of propulsion. Photon rockets with their enormous exhaust velocity were tempting, but not realistic. NERVA, NASA's nuclear solid-core engine (see Chapter 5), was considered, but its limited exhaust velocity of only 10 kilometers (6 miles) a second, would result in an impossibly heavy fuel supply. In the group's opinion, the fusion engine would be much too heavy and furnish very low acceleration. That left only the interstellar ramjet and the nuclear pulsed rocket, successor to Project Orion. Dr. A. Martin regarded the interstellar ramjet as too far beyond contemporary technology, and collecting huge masses of gas from the near vacuum of space did not appear practical. So by elimination, they decided upon the nuclear pulse rocket. The nuclear pulse engine would expel small spheres of frozen deuterium and helium-3 that would be exploded behind the ship by powerful electron beams. The gases expanding from the explosions would propel the ship to an ultimate velocity of more than 10,000 kilometers (6000 miles) a second, fast enough to reach Barnard's star in the allotted time.

At first a one-stage ship was planned. Mission time was set at 40 years to travel the 5.91 light-years to Barnard's star. Acceleration to cruise speed would take five years. The craft would attain 15 percent of light velocity and flash past Barnard's star at 44,600 kilometers (28,000 miles) a second. Decelerating was rejected, since it would demand a mass ratio of 8100 to 1!

The first attempt achieved a mass ratio of 150 for the ship. The engine would have an exhaust velocity of 10.7 million meters (32 million feet) a second and the payload would be 500 tons. At launch, the ship would have a mass of 100 million kilograms (100,000 tons) and be down to 1 million kilograms (1000 tons) when it passed Barnard's star. The enormous size of the probe is indicated by its reaction chamber, 100 meters (330 feet) in diameter. After a year of study, the group decided to redesign the ship to lower the mass ratio. This they did by extending the flight time and reworking Daedalus into a two-stage ship. The extension of flight time from 36 to 54 years and the shift to two stages reduced the mass ratio drastically, from 150 to 15.

The probe would have a mass of 47.6 million kilograms (104.8 million pounds) at launch, which might be from lunar orbit or from orbit around Jupiter. (See Figure 12-2, page 204.) Two days after launch, Daedalus would reach solar escape velocity; 250 days later it would cast off two of its first-stage fuel tanks, having passed 0.00037 light-year distance. At 0.0171 light-year it would cast off two more tanks and at 0.04888 light-year the last two. At 2.05 years into the mission, the entire first stage would be separated and the second stage would ignite. The second-stage engine would burn for 1.76 years, separating its four fuel tanks on the way once they were empty, until at 0.212 light-year distance from the Sun the engine would cease firing and the probe would coast at a constant speed of 12.8 percent of light velocity, 38,400 kilometers (23,800 miles) a second. This is equal to 138 million kilometers (85.6 million miles) an hour. We will discuss Project Daedalus further in connection with interstellar exploration (Chapter 15).

The Interstellar Ramjet

The BIS workers rejected the interstellar ramjet because it seemed too difficult. The situation is analogous to that of ocean shipping. From the fifteenth to the nineteenth century, Europeans traveled freely over the world's oceans in ships that needed no fuel, as they were propelled by the wind. The advent of the steamship ended that freedom. Coaling stations were built so that steamships could replenish their fuel supply. A new term, cruising range, had to be added to the maritime vocabulary. So too with starships. As long as they have to carry their own fuel supply, even the nearer stars will be barely in range at best.

Dr. Robert Bussard of Energy Resources Group put forward the idea of an interstellar ramjet in 1960.[21] He stressed the point that even fusion would be too weak to accelerate a starship to near light velocity. Instead of a starship that carried its own fuel, he advocated a vehicle that would use the interstellar matter diffused thinly through space and so have unlimited range and power to achieve near-light velocity. Interstellar gas would be swept into a nuclear reactor where it would fuse and propel the ship by acceleration of the exhaust or by projection of a photon beam. He calculated that if the ship had a mass of 1 million kilograms (2.2 million pounds) it would have to have an ion-collecting radius of nearly 60 kilometers (about 37 miles) and concluded:

> This is very large by ordinary standards but then, on any account, interstellar travel is inherently a rather grand undertaking, certainly many

magnitudes broader in scope and likewise more difficult than interplanetary travel in the solar system, for example. The engineering effort required for the achievement of successful short-time interstellar flight will likely be as much greater than that involved in interplanetary flight as the latter is more difficult than travel on the surface of the earth. However, the expansion of man's horizons will be proportionately greater and nothing worthwhile is ever achieved easily.

Bussard admitted that designing a working ramjet would be extremely onerous. Anthony Martin analyzed its problems in some detail.[22] Assume that the ship has a mass of 1 million kilograms (2.2 million pounds) and is to achieve an acceleration of 1 g (9.8 meters/sec^2 or 32 ft/sec^2). Even if it passes through a region of relatively high hydrogen density, a billion particles per cubic meter, it will have to have an intake radius of 60 kilometers (37 miles), while if it goes through a region of relatively low density, it must have an intake radius of 2000 kilometers (1200 miles). These are within the density limits of hydrogen clouds in space.

Obviously, a funnel 4000 km (2400 miles) or even 120 km (70 miles) across is beyond consideration. The intake has to be a magnetic field drawing in hydrogen ions from this huge volume of space. Neutral hydrogen would be ionized by laser beams. In effect, the starship would draw in the interstellar gas, boring a tunnel many kilometers wide as it sped through space.

Bussard's study was theoretical, while Martin attacked the technical problems. One of the most serious is that major losses will occur in mass and in radiation. As we have seen, fusion reactions transform only a small fraction of the mass into energy. In the case of the proton-proton reaction between hydrogen nuclei, it is only 0.0071 of their mass. For reactions between deuterium nuclei, it is 0.0009 or so, depending upon the reaction product. Too, only part of the hydrogen or deuterium will fuse. Radiation will also cause a serious loss of energy. The temperature in the fusion reactor will rise to hundreds of millions of degrees Kelvin, and much of the energy of the fusion reactions will appear as radiation, which will be useless for propelling the ship. For speeds over nine-tenths of light velocity, radiation losses will predominate.

The magnetic-intake fusion reactor will not be very different from the magnetic-confinement fusion reactor. Yet the ramjet and its reactor will be so much larger than any fusion reactor now even contemplated that the difficulties will be made even more severe by sheer scale. The electromagnetic coils generating the magnetic fields around the intake will be under great stress, since the magnetic fields will be much more intense than any produced to date. For example, the plasma in one of the Department of Energy experimental fusion reactors is confined by a magnetic field exerting on the plasma a pressure of 1.4 million newtons/m^2 (200 pounds/in.^{2}). The magnetic field in the ramjet intake may be a thousand times as strong.

As the speed of the ramjet rises, the magnetic flux density in the intake will have to increase accordingly to confine the increasing mass of the charged particles that are drawn in. The stress on the intake and on the ship itself will increase. Will the ship collapse? Preliminary calculations on this point have been reassuring. Oddly, if the support structure of the intake were made of diamond, a starship speeding though a low-density hydrogen cloud could accelerate at 1 *g* for 30 years and travel a distance of 3,800,000 light-years before the intake collapsed under the strain. This is nice to know, but how do you construct a diamond starship?

Drs. Gregory Matloff and Alphonsus Fennelly[23] take a more cheerful view of interstellar flight. (Matloff has also suggested, as we have noted, converting an O'Neill space colony into an interstellar space ark, a radically different proposition.) They believe that the scoop or intake can be used both to brake the ship and reduce the mass ratio. Consider a starship that carries 44 million kilograms (97 million pounds) of thermonuclear fuel and has a mass ratio of 10. The ship accelerates for 10 years at a maximum acceleration of 0.07 *g* to reach 12 percent of light velocity in traveling 0.7 light-years and consumes all its fuel. The ship slows down without expending fuel by unfurling a sail. The scoop gathers in interstellar hydrogen and blows it against the sail, which acts as a brake. Early supersonic airplanes deployed parachutes to slow down their landing. The sail does the same for the starship. Decelerating to 0.001 percent of the light velocity, 300 kilometers (186 miles) a second, takes about 20 years. However, since the hydrogen ions hit the sail at high speed, they heat it up to over 2000K. Since the sail is made of boron, it is safe. The sail is 10 kilometers (6 miles) in diameter and has a mass of 20,000 kilograms (44,000 pounds), although it is only 1/25,000 inch thick.

Later, Fennelly and Matloff came up with a new concept for a starship, the LINAC.[24] The novel feature of the LINAC is its superconducting wing batteries. The scoop is a cylinder 400 meters (1310 feet) long and 200 meters (650 feet) in diameter. The scoop, mounted on the fuselage of the starship, is coated with a superconducting niobium-tin coating. Currents in the scoop energize a vast magnetic field ahead of the ship, drawing in hydrogen ions from 10,000 km

(6200 miles) around, for in that region the ship's magnetic field is stronger than the interstellar magnetic field. The hydrogen is drawn into the ship's fusion reactor and then expelled through the cylindrical linear accelerator behind the scoop—hence the acronym, LINAC, Linear Accelerator Craft.

The wing batteries improve the ship's performance. Since the electrically charged wings are moving through the interstellar magnetic field, currents are induced across them, and the ship's kinetic energy (energy of motion) is converted into electricity. The electricity is used to accelerate the gases through the exhaust. If the fusion reaction is stopped and the electricity tapped, the wings act as brakes. The ship could be brought to a halt in one year from a speed of 30,000 kilometers (18,600 miles) a second. The boron-sail brake would take eight years. By drawing off the power in one wing, the ship could be gradually turned, just as a rowboat can be turned by rowing with one oar. The currents could also be passed through superconducting magnetic coils to generate the strong magnetic field needed for a magnetic-confinement reactor. If the reactor should break down, the wings could supply about a billion kilowatts to keep the ship's systems running!

While Fennelly and Matloff did not work out their design in as much detail as the Daedalus group, they furnished enough for a sketchy model. Imagine a LINAC 1216 meters (4000 feet) in length with wings each 304 meters (1000 feet) across and a mass of 18 million kilograms or 40 million pounds. LINAC could travel to Alpha Centauri at 0.3 peak light velocity and reach it in about 19 years.

Catalytic Nuclear Ramjet

Another block to interstellar flight is the composition of the interstellar medium. Fusion research has concentrated on the deuterium-tritium reaction, since it has the lowest ignition temperature of any of the possible fusion reactions, only 100 million degrees Kelvin. Unfortunately, almost all of the interstellar gas is hydrogen with only 1/100,000th deuterium. If the starship employs the deuterium-deuterium reaction, its scoop will have to draw in 100,000 times as much gas as it would if it relied upon hydrogen. If it uses the ionized hydrogen, or proton-proton reaction, the power of the reaction will be only one-billion-billionth that of the deuterium-deuterium reaction!

Do the scarcity of deuterium and the low reaction rate of hydrogen make interstellar flight impossible? Dr. Daniel P. Whitmire of the University of Southwestern Louisiana[25] has suggested a way out of this dilemma: have the reactor run on a catalytic nuclear reaction cycle such as the CNO bi-cycle that occurs in the depths of hot main-sequence stars. The cycle runs as follows.

$$^{12}C + p \rightarrow {}^{13}N + \gamma$$

$$^{13}N + p \rightarrow {}^{14}O + \gamma$$

$$^{14}O \rightarrow {}^{14}N + e^+ + \nu \quad e^+ + e^- \rightarrow 2\gamma$$ (This says that when the positron meets an electron, they annihilate, and give off two gamma rays.)

$$^{14}N + p \rightarrow {}^{15}O + \gamma$$

$$^{15}O \rightarrow {}^{15}N + e^+ + \nu$$

$$^{15}N + p \rightarrow {}^{12}C + \alpha$$

where:

- γ = gamma ray (gamma rays are shorter and more intense than X rays)
- p = proton (nucleus of a hydrogen atom)
- ν = neutrino (a particle that has no charge or mass)
- α = alpha particle (nucleus of a helium atom)
- e^+ = positron (nuclear particle with the mass of an electron but a positive instead of a negative charge)
- e^- = electron (smallest known particle with a negative charge; atoms are composed of a nucleus and a number of orbiting electrons)

The basic idea is to use the hydrogen, the bulk of the interstellar gas, to produce enough energy to drive the ship to near-light velocity. The ship then need not carry along most of its fuel supply. Notice that four protons (hydrogen ions) are used up in the cycle, while very energetic gamma rays are produced, together with neutrinos. The cycle repeats continually, since each of the isotopes in it acts as a catalyst; that is, each reaction keeps the cycle going. Any of the isotopes can be used to start the cycle. Get the cycle going, keep feeding in ionized interstellar hydrogen, keep the temperature and other conditions right, and the cycle will run by itself. No catalytic material will be drawn from the starship's fuel tanks. Hydrogen fuel will be needed to accelerate the starship to ramjet velocity but not thereafter.

Will the CNO bi-cycle do the job? Here we come to another

obstacle. The rate of energy production of a nuclear cycle is controlled by the rate of the slowest of its reactions. How fast is the CNO? It turns out that this cycle is 10^{18} times as fast as the proton-proton reaction. Fine, but is this enough to drive the starship to near-light speed?

Whitmire proposed a starship with a mass of 1.1 million kilograms (1100 tons) that would accelerate at 1 g. (See Figure 12-3, page 204.) Its propulsion system would convert into energy 0.67 percent of the hydrogen it collected and use it with 100 percent efficiency to reach 99 percent of light speed. At this speed, ship time is contracted to a tenth of Earth time. The power generated is 120,000,000,000 megawatts. The CNO bi-cycle reactor would run at a temperature of a billion degrees Kelvin and have a diameter of 36 meters (120 feet), big but feasible. Also, Whitmire pointed out, there is another catalytic nuclear cycle, the neon-sodium cycle, that would require a reactor only 19 meters (63 feet) in diameter.

The following table compares the specifications for Whitmire's neon-sodium bi-cycle reactor[25] with the present level of fusion technology. The specifications for the CNO bi-cycle reactor are the same except for the size of the reactor:

	Present Technology	Whitmire Reactor
TEMPERATURE	100,000,000K	1,000,000,000K
NUMBER DENSITY OF PLASMA	$1 \times 10^{12}/cm^3$	$1 \times 10^{19}/cm^3$
STRENGTH OF MAGNETIC FIELD	2.54×10^5 gauss	1.8×10^7 gauss

Another problem arises at one billion degrees, where radiation losses exceed the energy produced. The losses are due to fast electrons, and they may be overcome if at least part of the radiation is absorbed and converted into useful energy. Whitmire concluded, "The important point is that the catalytic proton burning model is at least theoretically capable of using the interstellar hydrogen to generate the power required for acceleration at 1 g to relativistic velocities."

Annihilation-powered Starship

Can science go beyond fusion and convert matter entirely into energy? This is theoretically possible, as Einstein showed many years ago, but is it even remotely feasible?

The annihilation of matter by antimatter would be the most powerful reaction ever known. For each kilogram of matter consumed, an annihilation reactor could theoretically yield 100 times as much energy

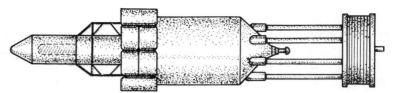

Figure 12-1. Orion.

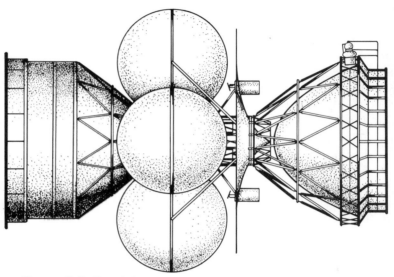

Figure 12-2. Daedalus.

Figure 12-3. Whitmire Catalytic Nuclear Ramjet.

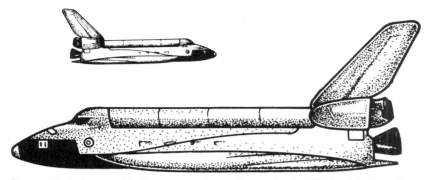

Figure 12-4. Space Shuttle, shown here for scale, is 123 feet in length.

as a fusion reactor. Since it would not violate the laws of nature, an annihilation reactor is possible—but can one be built? Dr. Gary Steigman, now at the Bartol Research Foundation, does not think so.[26]

To begin with, antimatter would have to be created, since in all probability there is virtually no antimatter in the solar system or in the Galaxy. If an antimatter particle meets its matter twin, they convert their energy into high-energy photons in an explosion. The Moon is ordinary matter, since none of the Apollo landing modules, Surveyors, Rangers, or the Russian Lunakhods exploded on touching the Moon's surface. Mars is ordinary matter, as shown by the Viking landings, Venus by the Venera landings, and Jupiter by the close passage of Pioneer 10. Meteorites are ordinary matter. As for possible antimatter in the Galaxy, Steigman believes that it would encounter ordinary matter in some form in about 100 years on the average.

Creating antimatter would consume more energy than would be generated. Antimatter generators would have to be developed, for atomic accelerators are not designed to produce antimatter efficiently, even in tiny amounts. An atomic accelerator yields only one antiproton for each thousand protons fired at the target. The process is so inefficient because the antiprotons shoot out of the target in all directions. Most of them evade the magnetic field, which is not designed to guide them into storage. Any antiproton that hits the inner wall of the accelerator disappears in a flash. If the target is too small, many of the protons will miss it. If it is too big, many antiprotons will hit the target and vanish. In a way, matter-antimatter works too well to be tamed. For example, the Los Alamos Linear Accelerator, with a 30-billion-electron-volt beam, would require about 200 million years to produce one kilogram of antiprotons!

Should we then dismiss the possibility of developing annihilation reactors? Steigman thinks so. Certainly, the annihilation reactor would demand an advance in technology as far beyond the fusion reactor as the fusion reactor will be from the first atomic pile. Significantly, however, NASA has not regarded the possibility as utterly hopeless. The Jet Propulsion Laboratory has examined annihilation for space propulsion. If an annihilation reactor could be constructed, mankind would have almost unlimited power at its disposal.

The Jet Propulsion Laboratory had two studies done by a consultant theoretical physicist, Dr. David L. Morgan. Morgan drew some interesting conclusions.[27] As a prototype concept, he sketched an annihilation-powered spaceship that would be employed in interplanetary travel. He began his analysis by assuming that antimatter could be produced in adequate quantity—which he believes impossible or extremely expensive with present-day technology—and that antihydrogen would be the easiest antielement to produce. The antihydrogen would be stored as a frozen sphere about 3 centimeters (1 inch) in diameter at near absolute zero in a high vacuum, suspended between two curved electrodes that would generate a strong electric field of up to 300,000 volts. Microwaves would sense the precise location of the sphere and automatically control the electrical charge to bring it back to its central position should it move from it by even a fraction of a millimeter.

The problem of drawing off a little antihydrogen at a time is attacked ingeniously. Small sectors of the sphere are irradiated with ultraviolet light, and antiprotons and positrons (antielectrons) are emitted. The antiprotons are drawn off through a hole in the upper electrode, while the positrons are drawn off through a hole in the lower electrode. Strong magnetic fields guide the stream of particles. The antiprotons are guided into the combusion chamber of the rocket motor which contains a magnetically confined plasma of a heavy element. The antiprotons annihilate in the nuclei of the heavy element, producing a huge release of energy along with pions and various nuclear fragments. The heated plasma is then released into the exit chamber of the rocket motor, where it is directed rearward from the spacecraft by a conical magnetic field that opens into space.

Morgan also developed a concept for an annihilation rocket motor of an interstellar spacecraft. In this case, the combustion chamber is eliminated and the antiprotons enter directly into the exit chamber. Here they annihilate with the protons in a jet of hydrogen, producing three or more very energetic pions or other particles for each antiproton. Most of the particles have an electric charge and are directed into space by the magnetic field.

The rocket exhaust of charged particles moves with nearly the velocity of light. This is desirable in an interstellar spacecraft, which attains a similarly high maximum velocity. For interplanetary spacecraft a much lower maximum velocity is required. Greater efficiency is obtained in this case by sharing the annihilation energy with the atoms and nuclear fragments of the heavy element. This produces a more massive exhaust with a lower velocity. In both cases, it is Morgan's opinion that roughly 50% of the annihilation energy could be converted into thrust.

The interplanetary spaceship would have a thrust of about 556,000 newtons (125,000 pounds) and reach a speed of 140 kilometers (86 miles) a second. The spaceship would be very small, having a mass of 18,000 kilograms (40,000 pounds) at takeoff—4500 kilograms (10,000 pounds) of payload, and 9000 kilograms (20,000 pounds) of inert propellant. It would accelerate at 3 g. For fuel, it would carry 1 gram (1/28 ounce) of antihydrogen!

Morgan's interplanetary spaceship, if realized, could be easily scaled up to a starship version.

The starship is a Mount Everest of technology. If its formidable challenge can be met—the unbelievable distances, the incredible energy requirements, the use of the ultrathin gas of interstellar space—then the human race will not only explore the stars: it will achieve a huge technological and scientific advance.

We have investigated in this chapter various ways to provide the propulsion needed. In the coming decades, advances in physics, particularly in fusion, should open the path to the stars.

Notes

1. Edward Purcell, "Radio Astronomy and Communication Through Space," USAEC Report, BNL-658, 1961.
2. Sebastian von Hoerner, "The General Limits of Space Travel," *Science*, 137 (July 6, 1962), 18. Copyright © 1962 by the American Association for the Advancement of Science.
3. Ernst J. Öpik, "Is Interstellar Travel Possible?" *Irish Astronomical Journal*, 6:8 (December 1964), p. 299.
4. Bernard M. Oliver, ed., *Project Cyclops–A Design Study of a System for Detecting Intelligent Life*, NASA/Ames Research Center, Code LT, Moffett Field, Calif. 94035 (CR 11455, 1971), p. 33.
5. Arthur C. Clarke, "Clarke's Third Law of UFOs," *Science*, 143

(January 16, 1968), 255. Copyright © 1968 by the American Association for the Advancement of Science.

6. Maxwell M. Hunter, "Accessible Regions Beyond the Solar System," AAS-69-386, *Advances in the Solar System*, 26 (1970), 308.

7. Gregory L. Matloff, "Utilization of O'Neill's Model Lagrange Point Colony as an Interstellar Ark," *Journal of the British Interplanetary Society*, 29 (1976), 755.

8. Ormond G. Mitchell, "Human Hibernation and Space Science," *Advances in Space Science and Technology*, 11 (1972), 249–265.

9. Freeman J. Dyson, "Death of a Project," *Science*, 149:3680 (July 9, 1965), 141.

10. Sebastian von Hoerner, "Population Explosion and Interstellar Expansion," *Journal of the British Interplanetary Society*, 28 (1975), 701.

11. Freeman J. Dyson, "Interstellar Transport," *Physics Today*, October 1968, p. 41.

12. Robert L. Forward, Senior Scientist, Hughes Research Laboratories, private communication.

13. G. Marx, "Interstellar Vehicle Propelled by Terrestrial Laser Beam," *Nature*, 211: 5044 (July 2, 1966), 23.

14. J. L. Redding, "Interstellar Vehicle Propelled by Terrestrial Laser Beam," *Nature*, 212 (February 11, 1967), 589.

15. Ernest Stuhlinger, "Photon Rocket Propulsion," *Astronautics*, October 1959, p. 36.

16. Krafft A. Ehricke, "Saturn-Jupiter Rebound—A Method of High-Speed Ejection from the Solar System," *Journal of the British Interplanetary Society*, 25:10 (October 1972), 561.

17. Dwain A. Spencer, "Fusion Propulsion for Interstellar Missions," *Annals of the New York Academy of Science*, 140 (December 1966), 407.

18. R. Hyde, L. Wood, and J. Nuckolls, "Prospects for Rocket Propulsion with Laser-Induced Fusion Micropellets," AIAA Paper 72-1063, November 1972.

19. Thomas A. Heppenheimer, "Some Advanced Applications of a 1-Million-Second I_{sp} Rocket Engine," *Journal of the British Interplanetary Society*, 28 (1975), 175.

20. Anthony R. Martin, ed., *Project Daedalus, Final Report*, (London: British Interplanetary Society, 1978).

21. Robert W. Bussard, "Galactic Matter and Interstellar Flight," As-

tronautica Acta, 6:4 (1960), 179. Copyright © 1960, Pergamon Press, Ltd. Reprinted with permission.

22. Anthony R. Martin, "Some Limitations of the Interstellar Ramjet," *Spaceflight*, 14:1 (January 1972), 21.

23. Gregory L. Matloff and Alphonsus J. Fennelly, "A Superconducting Ion Scoop and Its Application to Interstellar Flight," *Journal of the British Interplanetary Society*, 27 (1974), 663–73.

24. Alphonsus J. Fennelly and Gregory L. Matloff, "A Magnetic Interstellar Spacecraft," DH-7, March 1974 Meeting of the American Physical Society, American Institute of Physics.

25. Daniel P. Whitmire, "Relativistic Spaceflight and the Catalytic Nuclear Ramjet," *Acta Astronautica*, 2 (1975), 497. Copyright © 1975, Pergamon Press, Ltd. Reprinted with permission.

26. Gary Steigman, *On the Feasibility of a Matter-Antimatter Reactor Propulsion Sytem* (Yale University Observatory, 1973).

27. David L. Morgan, Jr., Consultant to the Jet Propulsion Laboratory, California Institute of Technology, private communication.

13 INTERSTELLAR NAVIGATION

If humans are to explore the Milky Way Galaxy, we must be able to determine our location and motion with great accuracy. We will need a knowledge of distances and an ability to follow a flight path precisely from origin to destination. This can be achieved by extending existing space navigation techniques. David G. Hoag and Walter Wrigley[1] of the Draper Laboratory claim that we are far closer to being able to provide interstellar navigation and guidance than to most other aspects of galactic exploration. They define *navigation* as the process of making the necessary measurements and performing the calculations to determine the position and velocity of the spaceship. *Guidance* is the process of measuring the spaceship's propulsion force and directing its magnitude, direction, and duration so the ship will follow the desired course.

In Chapter 7 we described coordinate systems that are suitable for the solar system, both for navigation and for locating places on planets and moons. For interstellar exploration, we will establish coordinate systems for all the bodies we will explore by use of these latter methods. For interstellar navigation, however, other coordinate systems must be used.

Galactic Coordinates

The galactic coordinate system (see Figure 13-1) is already being used to determine the structure of our Milky Way Galaxy. Its point of origin is the Sun. The mean plane of the Milky Way Galaxy is called the galactic equator. Its projection upon the celestial sphere is equivalent to the celestial equator in the equatorial system. The galactic equator passes along the central line of the Milky Way and serves as its reference plane. Galactic latitudes (b) are measured by using great circles that pass through the galactic poles, which are 90 degrees from the

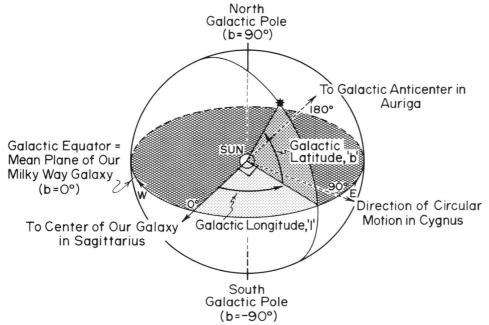

Figure 13-1. Galactic coordinate system. This is similar to the Earth-based coordinate system (see Figure 7-3). Galactic poles correspond to the Earth-based poles, galactic longitude to longitude, and galactic latitude to latitude. The directions for galactic rectangular coordinates are indicated (see Figure 13-2). **x** is toward the center of our Galaxy, **y** toward the direction of circular motion in Cygnus, and **z** towards the North Galactic Pole.

galactic equator. The North Galactic Pole is at 90 degrees, the galactic equator is at 0 degrees, and the South Galactic Pole is at −90 degrees. Although the Sun revolves about the Galactic Center in the opposite direction to the Earth's revolution around the Sun, the North Galactic Pole is, by convention, in the same hemisphere as the North Celestial Pole. Galactic longitudes (l) are measured along the galactic equator to the east from the direction of the Galactic Center (zero galactic longitude), which is in the constellation of Sagittarius.

The rectangular coordinate system, alternate to the spherical galactic coordinate system, has three mutually perpendicular axes labeled x, y, and z. Its origin is also at the Sun. The x-axis points to the Galactic Center. The y-axis, which is perpendicular to the x-axis in the galactic plane, points in the direction that the stars in the solar vicinity are moving around the Galactic Center. The z-axis points at the Galactic North Pole. (See Figure 13-2.) This rectangular system, which is suitable for interstellar navigation, is now used to study the motions and distribution of stars in the Sun's neighborhood.

A coordinate system must have reference points or a reference plane to establish the fundamental directions. It must also have objects whose positions are fixed with respect to each other so that we can determine the location of these reference points or planes and measure the changes in position of relatively nearby objects. Once the positions

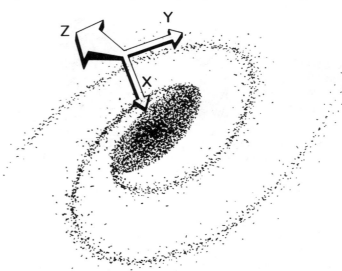

Figure 13-2. Rectangular galactic coordinates in relation to the Galaxy. **x** is toward the center of the Galaxy, **y** the direction of motion of a star at the Sun's distance, in circular orbit, and **z** the direction of the North Galactic Pole.

of these fixed objects have been accurately determined in any coordinate system, they can be used for alignment in that or any other coordinate system. For example, the locations of the celestial poles are found by Earth-based observers by examining the daily paths of stars whose declinations or celestial latitudes are nearly 90 degrees or -90 degrees.

Astronomers establish fundamental reference positions by using both distant stars and galaxies as fixed objects. The observed positions of the stars in the Milky Way Galaxy change slowly with respect to one another, because of their motions with respect to the Sun. This is particularly noticeable if they are astronomically near to the solar system. Most galaxies, on the other hand, are so distant that their motion through space cannot be determined and their relative directions are fixed, but their images are extended. This disadvantage will diminish to some extent once the pointlike images of stars are resolved into disks.

The stars do not seem to change their positions on the celestial sphere as the Earth travels around the Sun. Nor will they do so for astronauts traveling from the Earth to Pluto. The constellations—the patterns made by stars on the celestial sphere—look the same whether we observe them from the Earth or from Pluto, although their stellar positions show minute changes, which are clues to their distances. Hence, either stars or galaxies can be used to define the fundamental frame of reference when we are within the solar system. However, when we travel from the solar system to the nearby stars, the nearest stars will change their relative positions substantially. (See Fig. 13-3 and 13-4, pp. 214-15.) Therefore, for interstellar flight, the fundamental directions are likely to be defined in terms of distant galaxies.

Besides changing their relative positions, many nearby stars will change in brightness. A star gets brighter as we approach it and dimmer as we go away from it. Also, its apparent angular size depends upon the distance. If our velocity approaches the speed of light, these effects will be modified by the effects of relativity, which we will discuss later in this chapter.

Distance Determination

In conventional astronomical applications, the distances to bodies are very large compared to the longest available base lines. The triangles used to determine distances are very skinny. (See Figure 13-5.) The

Figure 13-3. Map of the sky as seen from the Solar System. The larger the symbol, the brighter the star.

sides AC and BC are almost the same as the distance to the object. Should we measure angles BAC and ABC at the ends of the base line or angle ACB at the apex? Consider two stations on Earth, A and B. To measure angles BAC and ABC requries that we determine the direction of A to B or B to A very accurately, which is extremely difficult. On the other hand, the parallax angle, angle ACB, can be measured by using the starry background. We measure the position of C on the celestial sphere from both A and B. Then we deduce the distance that C must be from the center of the baseline AB for the measured value of angle ACB to be at the apex of the triangle. Ideally, the base line should be perpendicular to the line connecting the center of the base line with C.

The parallax can be measured only as accurately as the direction to C can be found by using the background of stars on the celestial sphere. Under the most favorable circumstances, Earth-based observers can measure parallaxes to within a few thousandths of a second of arc by use of conventional techniques. This requires many measurements of relative positions. Parallaxes decrease as distance increases, so there is a distance limit to which one can measure a parallax within a given accuracy for a given base line. The longer the base line, the greater the limit. For Earth-based observers, the longest base line is the diameter of the Earth's orbit.

As the Earth moves around the Sun, the place from which we observe a star continually changes. The positions of nearby stars relative to those farther away also vary very slightly. If a star is in the plane of the Earth's orbit, its position shifts back and forth in a straight line on the celestial sphere as the Earth moves around the Sun. A star 90 degrees away from the plane of the Earth's orbit describes a circle, while stars in between describe ellipses. The angular size of one-half the largest diameter of the star's parallactic ellipse is called the *stellar parallax*. From the star's point of view, it is the angle subtended at the star's distance by one astronomical unit, the mean distance of the Earth from the Sun.

Measured parallaxes must have small corrections applied to account for the parallaxes of more distant stars. Besides, the stars move relative to the Sun. To separate the parallaxes from the star's motion, the star is observed on the same date of the year for several years. This

Figure 13-4. Map of the sky as seen from possible planets around Tau Ceti. The Sun is marked by the hourglass symbol near 14 hours right ascension. The orientation of coordinates is the same as in Figure 13-3.

insures that the Earth is at the same place in its orbit when the star is observed. Any change in the star's position must then be caused by its motion relative to the Sun.

The geometry of parallax measurements defines a new unit of distance, the parsec. The circumference of a circle is $2\pi r$, where r is the radius. In Figure 13-5, A and B are points from which the direction to the star C is measured and p is the parallax. If the baseline AB has a fixed length, then, as r becomes longer, both AC and BC approach R in length. AB is in the same ratio to the circumference of a circle of radius r as the angle p is to 360 degrees. Thus:

$$\frac{AB}{2\pi r} = \frac{p}{360°} \qquad (13\text{-}1)$$

A circle contains 360 degrees, each degree is 60 minutes of arc, and each minute is 60 seconds of arc. When we solve equation (13-1) for r and convert 360 degrees into seconds of arc, then

$$r = 206{,}265 \, \frac{AB}{p \text{ (seconds of arc)}} \qquad (13\text{-}2)$$

The parallax is defined as the angle, measured in seconds of arc, that is subtended by one astronomical unit at the star's distance from the Sun. The distance of the star in astronomical units is therefore $206{,}265/p$. Astronomers designate the distance, 206,265 astronomical units, as one *parsec*, abbreviated *pc*.

Since 1 AU is 1.496×10^8 km, 1 pc = 3.086×10^{13} km. One parsec corresponds to the distance at which a star would have a parallax of one second of arc. Also, the distance of a star in parsecs is

$$r = \frac{1}{p} \qquad (13\text{-}3)$$

where p is the stellar parallax in seconds of arc. For example, a star whose parallax is 0.1 second of arc is 10 pc away.

Another unit of distance, which we mentioned earlier, is the light-year (LY), the distance traversed in one year by photons, particles of light, in a vacuum. Since the velocity of light is 2.998×10^5 kilometers per second, and there are 3.16×10^7 seconds in a year, 1 LY is 9.46×10^{12} kilometers. One parsec therefore is equal to 3.262 LY. Astronomers use both units in measuring distances. The parsec, as we have seen, has a geometrical meaning, while the light-year has a physical one, as it is the greatest distance that a signal can travel in a year.

Figure 13-5. Parallax determination. The star at C is at a distance γ from A and B. The baseline AB subtends an angle ρ.

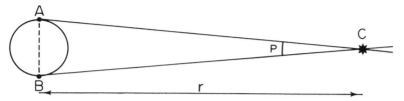

The Sun's nearest neighbor is the Alpha Centauri triple star system. Alpha Centauri A and Alpha Centauri B are a close binary system, while Proxima Centauri orbits the pair in a relatively large orbit. With their parallax of 0."76 (0.76 arc-seconds), they are 1.3 pc (4.3 LY) distant. Alpha Centauri is too far south to be seen in most of the United States. From most of the Northern Hemisphere, the nearest visible naked-eye star is Sirius, 2.6 pc (8 LY) distant.

Errors in parallax are typically a few thousandths of a second of arc as measured from the Earth's surface. Stellar parallaxes have to be larger than 0."05 to have errors of less than 10 percent. Within the radius of 20 parsecs from the Sun, there are slightly more than 1000 known stars whose parallaxes can be refined to this limit.

Astronomers have devised many methods for obtaining stellar distances, all of them depending directly or indirectly upon stellar parallaxes. Observing stars from space will greatly extend the usefulness of this basic method. Interstellar travel will lengthen the baseline enough so the distance to nearby stars will be determined by rigorously solving the distance triangle instead of using the skinny triangle approximation. The formula that we derived for the parallax will not be appropriate. Rather, the triangle will be solved by measuring the two angles at the end of the base line, and determining the length of the baseline.

Fortunately, the base line made by the trajectory of a starship on an interstellar mission will be far longer than 2 AU. Let us call the target star *a* New and let *b* Unk be the star whose distance is to be determined. The angles *b* Unk–Sun–*a* New and *b* Unk–*a* New–Sun can be precisely derived from measurements on the celestial sphere. (See Figure 13-6.) The first angle, *b* Unk–Sun–*a* New, is found from determinations made in the solar system; the second, *b* Unk–*a* New–Sun, from those made in the vicinity of *a* New. Then we can calculate the distances of Sun–b Unk and b Unk–a New.

For distance determinations, we are concerned with seeing, which is made problematic by the unsteadiness of the atmosphere. Since most stars have too small an angular size for our existing telescopes to resolve their disks, their images should be very small diffraction patterns, whose centers look like tiny circles. However, when the seeing is bad, a star appears as a fuzzy circle that dances around because of the distorting effects of the Earth's atmosphere. Under typical atmospheric conditions, a stellar image has an apparent diameter of 1 to 4 seconds. Under unusually excellent conditions, stellar images have diameters of about ¼ second, which is approximately the angle that can be resolved by a 50-centimeter (20-inch) telescope.

Astronomers measure the parallax of a star by finding the angular distance between the centers of stellar images. This process requires a series of measurements. The best ground-based analyses have errors of 0".005, which implies 10 percent accuracy out to 20 pc. The recent introduction of computer techniques that can handle a greater number of measurements between stellar images should reduce the errors to about 0".003, or, equivalently, increase the distance for 10 percent accuracy to 30 pc.

In space, the images of the stars will appear smaller than they do in a similar telescope on Earth. Using the Space Telescope, which is

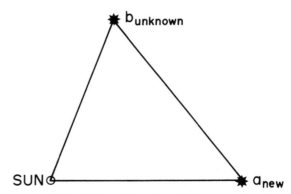

Figure 13-6. Determining the distance of a star from a starship. The Sun, a_{new} and $b_{unknown}$. In interstellar travel, we will no longer use skinny triangles but instead solve the usual triangles which we learned in geometry. The angle $b_{unknown}$-Sun-a_{new} is measured from the solar system while the angle Sun-a_{new}-$b_{unknown}$ is measured from the viewpoint of a_{new}.

stabilized in pointing to 0".007, astronomers in the 1980s will be able to determine the angle between the centers of two stellar images to 0".002 in a single measurement. Series of observations will permit reduction of this uncertainty to 0".0005, or 10 percent accuracy out to 200 pc. Increasing the size of the main mirror from 2 meters to, say, 25 meters (984 inches), the proposed size of the next generation of large telescopes, should yield the accuracy of 0".00004, or 10 percent accuracy out to 2500 pc. But 2500 pc is not very far on the galactic scale, as it is only 30 percent of the distance from the Sun to the Galactic Center.

If a 25-meter (984-inch) telescope were located on Earth instead of in space, its ability to determine parallax would be degraded. There is a new technique, speckle interferometry, that circumvents the turbulence of the atmosphere. The full angular resolution of the telescope can be achieved, but there is a loss in limiting magnitude. For a 25-meter telescope operating in visible light, a resolution of 0".003 is predicted. By repeated measurements, astronomers should be able to determine the angle between the centers of two stellar images to a twentieth of this or less, say, 0".00015, which gives at least 10 percent accuracy out to 650 pc, or 0.1 percent at 6.5 pc. As of this date, however, no parallax measurements using speckle interferometry have been reported because of the relatively narrow field of view of existing speckle cameras. (See also Chapter 14.)

Angular resolution can be improved by building larger telescopes. But there are other ways to improve parallax measurements by use of space telescopes. The surfaces of the mirrors and other optical components of a telescope are within tolerance limits of their ideal figures. When the differences between the real and ideal surfaces are small enough, a telescope will function in a diffraction-limited mode. That is, it will give the smallest possible image at each wavelength longer than a given wavelength. Shortward of this given wavelength, the image size will remain constant. The planned limiting wavelength of the Space Telescope is 6325 angstroms, which is in the red region of the visual spectrum. Smaller images could be obtained by reducing this wavelength, but eventually we will come to some limiting size for a single instrument. On Earth, atmospheric turbulence spreads the image so that its size is greater than the diffraction-limited one.

Let us consider what can be done by increasing the length of the base line and using a single 25-meter telescope or a pair of them in space. If larger telescopes are developed, then the potential improvement in parallax measurements will be so much the greater. Our first choice might be to place a telescope in orbit around Mars. Since Mars is 1.5 AU from the Sun, this increases the base line by 50 percent. Jupiter, 5.2 AU from the Sun, would be an even better location, since it would increase the baseline by over 500 percent. This base line would yield 10 percent accuracy out to 12,500 pc. However, the need to separate the stellar motions from the parallaxes requires observations that would take three or four revolutions of Jupiter around the Sun or 36 to 47 years.

To reduce the waiting time, we could pair telescopes in orbit around the Earth and Jupiter, or Mars and Jupiter, or on an asteroid and about Jupiter, and make simultaneous measurements. This would eliminate the need to make repeated measurements at a given place in the orbit for several revolutions in order to separate the stellar motion from the parallax. Even better, we could put a pair of telescopes at Pluto's distance from the Sun (40 AU) at opposite points of its orbit. In 125 years, half of Pluto's period of revolution, astronomers could measure distances to 100,000 pc with 10 percent accuracy. We could then determine the distance of every observable star in the Milky Way Galaxy to at least 3 percent accuracy. This is equivalent to 0.1 percent accuracy at 1000 pc and 0.001 percent accuracy at 10 pc.

Another way to obtain such accuracy would be to send a spaceship to a nearby star. One 25-meter telescope is sent on the ship while its twin remains in the solar system. Let us assume a voyage out to 2 pc. If we are going to go that far, we might as well plan to map the entire celestial sphere.

Mapping the celestial sphere would require two starship expeditions and three telescopes. One of the starships would travel, say, toward the Galactic Center and the other toward one of the Galactic Poles. We would thereby extend our baseline 206,265 times the length of the one currently in use. Two such trips would yield parallax determinations with 10 percent accuracy out to at least 5.15×10^8 pc, which is equal to 1.68 billion light-years. For comparison, the distance of M31, the Great Spiral Galaxy in Andromeda, is 2.2 million light-years. The astronomers would have directly measured distances 750 times as far as M31. They would have determined the distances of all the galaxies in the local supercluster centered on the Virgo Cluster of Galaxies as well as of a substantial fraction of the universe.

Solar system and interstellar travel will increase our knowledge of stellar distances by lengthening the base line for parallaxes. Before going on a space voyage, either within the solar system or beyond it, we must accurately determine the distance to our destination. Our 25-

meter telescope in orbit around Jupiter gives an accuracy of 10 percent or better out to 12,500 pc. This is equivalent to determining the distance to the center of mass of the Alpha Centauri system to within 2.5 AU. For comparison, the distance between Alpha Centauri A and Alpha Centauri B, which form a close binary pair, is 25 AU.

Motion Measurements

The Sun's speed relative to the nearby stars is 4 AU a year, which is typical of the stars in the Sun's neighborhood. Suppose it took five years to reach Alpha Centauri and that Alpha Centauri were moving at the same speed as the Sun. If our navigator ignored the space motion of Alpha Centauri, he would miss it by about 20 AU. Astronomers have measured the relative speeds of the nearby stars within a margin of error of less than one AU per year and will do it better from space.

The positions of stars on the celestial sphere slowly change over the years. This stellar proper motion (see Figure 13-7, page 222) is measured in seconds of arc per year. Most stars have proper motions larger than their parallaxes, while only 330 stars are known to have proper motions greater than one second of arc per year. Barnard's star has the largest known proper motion, $10''\!.25$ per year, yet even at this rate it takes 350 years to move one degree on the celestial sphere. The mean proper motion of all the naked-eye stars is less than $0''\!.1$ per year. Astronomers give the proper motion in both right ascension and declination separately to specify both the direction and the movement of the star. Since telescopes in space will be able to measure smaller changes in the positions of the stars than their Earth-based counterparts, proper motions of the nearby stars will be measured with greater precision, and determining the proper motions of more distant stars will be possible.

Radial velocity is the component of the space velocity of the star along the line of sight (to and from us). It tells how fast a body is approaching or receding from the Sun. The astronomer determines radial velocity by measuring the shift of the lines in its spectrum. This can be done for any star or galaxy bright enough that a spectrogram can be taken. Radial velocity is usually stated in kilometers a second. It is denoted as positive if it is receding and negative if it is approaching. In our vicinity, as many stars are approaching as are receding, so

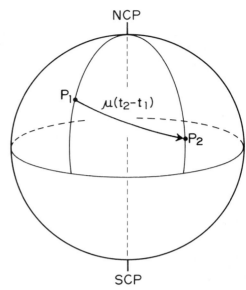

Figure 13-7. Proper motion on the celestial sphere. At the time t_1, the star is at P_1, and at the time t_2, at P_2. In the time interval $t_2 - t_1$, it moves μ $(t_1 - t_2)$ on the celestial sphere. μ is the proper motion, usually expressed in seconds of arc per year.

the solar neighborhood is neither expanding nor contracting. Kapteyn's star has the largest radial velocity of the stars in the solar neighborhood, 242 kilometers per second. The average absolute value of the radial velocity of local stars is 22 kilometers per second, while 15 kilometers per second is the most common value.

The other component of the space velocity is the *tangential velocity*, the motion across the line of sight or in the plane of the sky. The tangential velocity is calculated from the proper motion and the distance—quantities more difficult to measure than the radial velocity. Consider two stars, A and B, at the same distance but with different proper motions. (See Figure 13-8.) If A moves farther in a year than B, both its tangential velocity and proper motion will be greater. Consequently, tangential velocity is proportional to the proper motion. Now consider two other stars, C and D, which have the same proper motion while C is closer to the Sun than D. (See Figure 13-9.) Since D is farther away, it must travel a greater distance to describe angle μ, hence it has a greater tangential velocity. Tangential velocity is there-

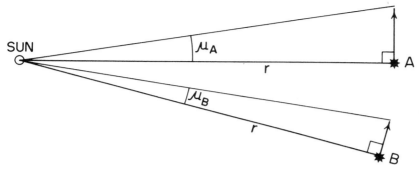

Figure 13-8. Proper motion and tangential velocity for stars at the same distance. Stars A and B are each at the distance **r** from the Sun. Their motion is perpendicular to their Sun-star directions. Since A has a larger tangential velocity, it moves farther in a year and its proper motion is greater than B's.

fore proportional to distance. These arguments show that the tangential velocity is proportional to both the distance and the proper motion. Numerically:

$$T \text{ (km/s)} = 4.74\mu \text{ (sec/yr)} \cdot r \text{ (pc)} \qquad (13\text{-}4)$$

where
- T = the tangential velocity
- μ = proper motion
- r = distance

The factor 4.74 is a numerical constant that converts kilometers per second into seconds of arc times parsecs per year. Kapteyn's star is moving with a tangential velocity of 163 kilometers per second, the fastest tangential velocity of the nearby stars. Typical tangential velocities of stars in the solar neighborhood are between 20 and 25 kilometers per second.

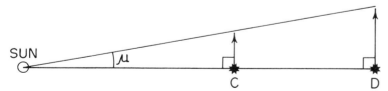

Figure 13-9. Stars with the same proper motion but different distances. Stars C and D have the same proper motion. However, since D is farther away than C, its tangential velocity is greater than C's.

To completely describe the motion of a star relative to the Sun, we must give both the direction in which it is moving and its speed. The radial and tangential velocities are the two components of the space velocity and they are at right angles to each other. The Pythagorean theorem tells us that

$$V = \sqrt{T^2 + V_r^2} \qquad (13\text{-}5)$$

where

V = space velocity,
T = tangential velocity,
V_r = radial velocity.

The direction is found by examining the components in three-dimensional space.

So far, we have considered the Sun to be at rest and have measured the movement of the stars with respect to the Sun. Now, we want to find the Sun's own motion. To simplify the problem, consider a space traveler and assume that he can ignore the motions of the stars as his starship travels from α to β. (See Figure 13-10). Stars 1 and 4, which are respectively straight ahead and behind the ship, do not change their directions, so their proper motions are zero. Stars 2, 3, 5, and 6 have proper motions directed behind the ship. The closer the stars, and the more nearly they are 90 degrees to the direction of motion of the ship, the greater their proper motion. Our space traveler could deduce his forward motion from the backward drift of the nearby stars.

Another way to observe the ship's motion is by observing the radial velocities of the stars. Star 1 would show a blue shift, while star 4 behind would exhibit a red shift. Any star ahead will show a blue shift, but when it is overtaken and at right angles to the ship's direction of motion, it will not show a shift. As it recedes, it will begin to show red shifts. In this way, we have two complementary methods for deducing the motion of the starship.

The stars behave toward the Sun in the same way. For the Sun, we treat the motions of the stars in a statistical manner. From both kinds of analyses, astronomers have found that the Sun is moving in the direction now occupied by Vega, the bright star in the constellation of Lyra, at the speed of 4.2 AU per year or 20 kilometers a second. Our space navigator, however, will have sufficient knowledge

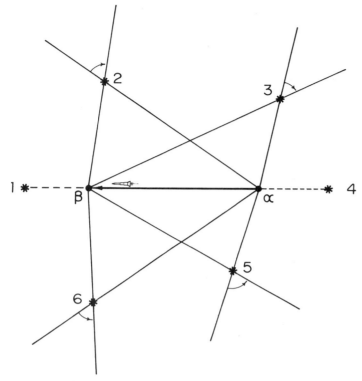

Figure 13-10. Apparent proper motion and Doppler shift. Assume that stars 1 to 6 are at rest. As the starship moves from α to β, stars 1 and 4 show no proper motion, while stars 2, 3, 5, and 6 apparently move backwards, due to the forward motion of the starship. Star 1 shows a blue shift while star 4 shows a red shift. Stars 2, 3, 5, and 6 show blue shifts while the starship is approaching them and red shifts while the starship moves away from them.

of the motions of the nearby stars so he will be able to make the corrections on an individual basis and so accurately determine the velocity of his starship.

As we have noted, the starship navigator will observe changes in the brightness of stars, the shapes of constellations, and the angular sizes of the stars' disks. His measurements of the radial velocities of the stars will reflect the ship's velocity as well as theirs.

Another phenomenon important to the starship navigator is the aberration of starlight, which shifts the position of all objects toward the observer's apex of motion, or the direction in which he is going. It applies equally well to all objects in a given direction. A person walking in a rainstorm in which the rain drops are falling vertically has the sensation that the drops are falling at an angle tilted forward from the vertical. In the same manner, a starship runs into the light. The Earth's orbital motion causes the stellar positions to move up to 20″.5 of arc from their mean values. (This is a complication that we ignored in our discussion of parallaxes.) The aberration of starlight depends only upon the velocity of light, the velocity of the starship, and the angle between the position of the star and that of the starship's course. Its effects can be calculated with great accuracy.

The appearance of the celestial sphere becomes very distorted when an appreciable fraction of the speed of light is reached. As the ship's speed increases, the star field compresses in the direction of the ship's motion. This compression is itself a source of information on the starship's speed. At $0.1c$, (c equals the velocity of light), the maximum aberration angle is about 5 degrees; at $0.3c$, 18 degrees; at $0.5c$, 31 degrees; at $0.7c$, 48 degrees; at $0.9c$, 77 degrees; while at $1.0c$, the speed of light, the entire field collapses to a point ahead.

Hoag and Wrigley[2] have suggested that interstellar navigation will be done primarily with inertial instruments, as is most Earthbound travel by airplanes and ships. Periodic measurements of the stars will be used to improve the accuracy of the current position and the velocity of the starship. In this way, the starship navigator will have a good estimate of the values that he wants to improve. The types of instruments used in inertial navigation are clocks, gyroscopes, and accelerometers. Clocks will be used for many other purposes besides measuring the passage of time. Gyroscopes sense angular motion, while accelerometers give the difference between gravitational and total acceleration. They find, for example, the acceleration caused by the firing of the starship engines. When the starship is approaching a solar system, once the position of the starship is found, the gravitational components of the total acceleration can be found by analysis.

We will need an Interstellar Flight Atlas to accurately determine where we are in space. The Atlas will be based upon data accumulated from years of measuring parallaxes and other properties of stars such as stellar diameters, velocities of rotation, energy distributions, masses, spectral types, companions, planetary systems, the orientation of their rotational axes in space, and any other information that uniquely identifies a star. We have shown how space astronomy will improve determinations of the distances of stars in the solar neighborhood. Space-based observatories will also greatly expand and improve our knowledge of the properties of nearby stars. The first interstellar flights will contribute even more accurate data.

Besides his inertial measurements, the interstellar navigator will have many methods for determining the position and velocity of his ship. He will use the Interstellar Flight Atlas as a major reference. He could proceed by a three-dimensional analog of Earth-based surveying that requires deducing the angles between three stars that do not lie on the same line and solving the resulting triangles. The computation leads to two possible positions of the ship, and a fourth reference star is needed to eliminate one possibility. Moskowitz and Devereux[3] have designed a hyperaccurate sextant for interstellar navigation based upon the design of the sextant used by the Apollo astronauts.

Other methods, especially near a star, would involve measuring the star's brightness or diameter or both. This would give the ship's distance from the star. The navigator would then determine the positions of three reference stars and solve two triangles involving the ship's location and the angles they make with the ship to complete the determination of the ship's location.

For determining the ship's velocity, the navigator could take two fixes some time apart and derive the average velocity by dividing the distance traveled by the time that elapsed. Alternatively, he could measure the Doppler shifts of at least three stars in widely different directions. By subtracting their space motions, he could deduce the starship's velocity. (This is much like the method used for determining the Sun's motion.) The navigator could also determine the starship's velocity by the communication signals received from the solar system.

Astronomers on Earth use the Doppler shifts of spectral lines to determine the stellar line-of-sight motions with respect to the Sun. As these motions are usually in the tens-of-kilometers-per-second range, their Dopplers shifts are tenths of angstroms. The velocities of interstellar flight will be much greater, and so will be their observed phenomena.

When the starship is traveling at $0.1c$ (30,000 kilometers or 18,600 miles a second), the Doppler shift amounts to about 10 percent of the rest wavelength (wavelength measured in the laboratory) along the direction of motion. In the forward direction, cool stars appear brighter and hot stars fainter, while in the backward direction, cool stars appear fainter and hot stars brighter. This is caused by the shift in wavelengths that correspond to the optical wavelengths we see from Earth and the shift of the peak of the stars' energy distribution. There is no effect at 90 degrees to the direction of the ship's flight. As the starship's speed increases to 0.3 or $0.6c$ (180,000 kilometers or 112,000 miles a second), these effects become more pronounced. Dr. James R. Wertz[4] has noted that blue and red supergiant stars would thus be very useful as galactic beacons for interstellar navigation. Also, these stars are very luminous and can be readily identified over a large range of ship velocity.

As the starship travels away from the Earth, it will take longer to receive messages, and the reverse will happen when the ship is homeward bound. There is another effect. When a clock is moving relative to an inertial system that has a second clock, its time passes more slowly than the time marked by the inertial-system clock. This is called *time dilation*. Consequently, a space traveler takes less time to make a voyage according to a clock that he takes with him than according to a clock he left at home.

Time-dilation effects begin to become significant at about $0.3c$ (90,000 kilometers or 56,000 miles a second), and they increase as the speed of light is approached. At $0.3c$ the traveler's time is slower by about 0.5 percent, while at $0.5c$ (150,000 kilometers or 93,000 miles a second) it is slowed by 14 percent compared with that of a body moving slowly. A 4.3-light-year trip from the solar system to Alpha Centauri would take 14.3 years at $0.3c$, while the crew on board the starship would find that it takes 13.7 years. This amounts to about a year's savings on a round trip to Alpha Centauri.

In brief, instead of being "lost among the stars," the starship's crew would know their location and velocity precisely at all times.

Notes

1. David G. Hoag and Walter Wrigley, "Navigation and Guidance in Interstellar Space," *Acta Astronautica*, 2 (1975), 513. Copyright © 1975, Pergamon Press, Ltd. Reprinted with permission.

2. Hoag and Wrigley, "Navigation and Guidance," p. 513.

3. Saul Moskowitz and William P. Devereux, "Navigational Aspects of Transtellar Space Flight," *Advances in Space Science and Technology*, 10 (1970), 75.

4. James R. Wertz, "Interstellar Navigation," *Spaceflight*, 14 (1975), 206.

Note: Another fine reference on interstellar navigation is G. R. Richards, "Project Daedalus: The Navigation Problem," *Journal of the British Interplanetary Society*, 28 (1975), 150.

14
DETECTION OF EXTRASOLAR PLANETS

We have only one example of a star with a planetary system: our Sun and its solar system. More than half the stars we can observe with a telescope are members of double or multiple star systems. The theory of star formation suggests that it is easier to produce stars with planets than isolated stars, and perhaps even necessary to do so. There are many reasons for believing that there are planets about other stars. The most important is simply that we have no reason to believe that the Sun is unique in this respect. The challenge is to observe these extrasolar planets.

Astrometric Search for Extrasolar Planets

Dr. Shiv Kumar[1] of the University of Virginia has used the theory of stellar interiors and the related nuclear physics to show that the least massive object that can be a true star is one with a mass of about 0.07

that of the Sun. To be a true star, it must attain a high enough central temperature to fuse hydrogen into helium. Those objects with masses of 0.07 to $0.01 M_\odot$ ($M_\odot$ equals the mass of the Sun) will shine like stars for millions of years while undergoing complete gravitational contraction on the way to becoming black dwarfs. If the object has a mass of $0.01 M_\odot$ or less, it will hardly glow at all in the optical region and thus will be considered a planet.[2] The least massive known star is probably the companion to Ross 614. It has a mass of $0.06 M_\odot$. Yet, Jupiter radiates two to three times as much energy as it receives from the Sun, primarily in the infrared.

Dr. Peter van de Kamp of Sproul Observatory has been one of the leading searchers for planetary companions of stars by the method of long-focus photometric astrometry as part of the more general problem of finding unseen companions of stars. His approach[3] is to study the perturbations of a star's motion as produced by its unseen companion. If such an object is present, the visible star will describe an orbit around the center of mass of the two objects. The path of the star against the background of fainter reference stars is deduced from photographs made with long-focus refractor telescopes. After corrections are made for the motion of the Earth, the deviations from a straight path in the sky are extremely minute. The threshold for discovery is about $0''.02$ or one micron (10^{-6} meter) as recorded at the focal plane of the telescope. Unless the star is relatively close and the secondary's period is fairly long, it is almost impossible with existing equipment and conventional techniques to detect companions of less than $0.01 M_\odot$ for stars of about one solar mass. The range of masses detectable for a hypothetical planet depend upon the mass of the primary star, the primary's distance from the Sun, the orbital period of the planet, as well as on the accuracy of the positional measurements and the orientation of orbit relative to the plane of the sky. Besides, the presence of other planets in the system may affect the detection of the most massive one.

At the limit of detection, many unseen objects are indicated, with masses of $0.06 M_\odot$ or so that are probably stars or proto-black dwarfs. The observations required to detect them take from several years to several decades. Once the proper motion, stellar parallax, and the aberration of starlight have been taken into account, one can look for systematic trends in the data—that is, oscillations in the predicted path that would suggest the presence of a companion. The great difficulty is that these systematic trends can be close to, or less than, the accuracy of the observations and could be due to some systematic or random error in the techniques of photographic astrometry.

Most of the recent interest in unseen companions has centered on Barnard's star, which is 6 light-years (1.8 pc) away and the second nearest known stellar system to the Sun. It is a member of the galactic halo population and has an orbit about the Galactic Center quite different from that of the Sun. It is now passing through the plane of our Milky Way Galaxy and is moving relatively fast compared to stars like the Sun. Barnard's star is spectral type M5, which implies that its mass is about $0.15 M_\odot$. It has a lower concentration of heavy-element atoms compared to hydrogen atoms than the Sun.

Let us suppose that Barnard's star has a planet. Its orbital period P, stellar mass M, planetary mass m, and the relative semimajor axis $A + a$ of the system are linked by the following version of Kepler's third law:*

$$(m + M)P^2 = (A + a)^3$$

A and a are the respective semimajor axes of the star and the planet with respect to their mutual center of mass. A and a are given in astronomical units, P is in years, and M and m are in solar masses. The definition of the center of mass tells us:

$$\frac{M}{m} = \frac{a}{A}$$

Unfortunately, only P is directly observable. A can be deduced from the perturbation, or deviation from a straight path, but there are still three unknowns, M, m, and a, and only two equations. To solve the problem, one of these three quantities has to be determined independently. The spectrum and absolute magnitude of the star can yield a reasonable value of M, the star's mass. So by measuring P and A, we can deduce m and a, the planet's values. If we could directly observe the planet, the measurement of these values would be far simpler and, more important, more certain.

Van de Kamp published in 1963[4] an analysis of 2413 plates taken mostly between 1937 and 1962 by fifty observers with the 61-centimeter (24-inch) refractor of Sproul Observatory. He concluded that the data showed deviations that could be produced by a companion with a mass of $0.0015 M_\odot$ or 1.6 times that of Jupiter in an orbit with a period of 24 years and a semimajor axis of 0″.024 with an eccentricity of 0.60. In 1969 he reported[5] that an additional five years of plate material and a

---

*The semimajor axis of an elliptical orbit is one-half its longest dimension.

reanalysis of his data generally confirmed his previous results. Later the same year[6] he showed that an alternate analysis gave two companions in nearly coplanar circular orbits with periods of 26 and 12 years and masses of 1.1 and 0.8 times that of Jupiter, respectively. Unfortunately, with the data he had, he could not decide between the results of the two analyses.

In 1973 three analyses of Barnard's star appeared, two of which reanalyzed van de Kamp's data while the third presented a completely different set of results. Oliver G. Jensen and Tadeusz Ulrych[7] of the University of British Columbia found by a power-law analysis that there may be three planets with masses similar to Jupiter's orbiting about Barnard's star. Drs. David C. Black and Graham C. J. Suffolk[8] of the NASA Ames Research Center found that at least two massive planets orbit this star in orbits that are inclined to one another by at least 50 degrees. Both analyses assumed that van de Kamp's data were free from error, as large-scale, uncorrected, systematic errors would probably invalidate their conclusions.

Drs. George Gatewood of Allegheny Observatory and Heinrich Eichhorn[9] of the University of South Florida analyzed the motion of Barnard's star from 241 plates taken at the Allegheny and Van Vleck Observatories. This material was not as uniformly distributed in time as the Sproul plates. Gatewood and Eichhorn found that van de Kamp's values for a single planet had moderate uncertainties associated with them, but they noted that this does not necessarily give a complete picture, as long-focus astrometric investigations can have hidden errors. These errors can be found only by comparing results obtained at two or more independent sources. Unfortunately, the plate material that Gatewood and Eichhorn studied did not confirm the presence of planetary companions. Since their reduction techniques were, in fact, somewhat different from those of van de Kamp, they suggested the possibility of a systematic error in the Sproul material. In 1976, van de Kamp[10] applied such a correction to the Sproul 1942–1948 material, which probably invalidates the models of Black and Suffolk and of Jensen and Ulrych.

Nevertheless, van de Kamp's reanalysis of these data in 1976, with the inclusion of the most recently obtained material, and the recognition that some of his earlier material may have been subjected to a systematic error, led him to conclude that there were two periodicities in the perturbation—that is, systematic deviations about the straight-line path—one of 11.7 years, which is well established, and a second of 18.5 years, which is more uncertain. With a mass of

$0.14 M_\odot$ for Barnard's star, the companions have masses of $0.00086 M_\odot$ and $0.00040 M_\odot$, with errors of about $0.0001 M\odot$. These results are not far above the threshold of attainable accuracy. The systematic errors may be comparable to, or even larger than, the amplitude of the perturbation.

Gatewood[11] in 1976 attempted to correct the Sproul material for the same systematic error as van de Kamp did later, and then to combine it with the Allegheny and Van Vleck material. His results suggest a gravitational interaction of Barnard's star with one or more Jovian-mass planets. But he did not derive parameters for this system, as he believed the results were still possibly subject to undetected systematic errors. We can conclude that Barnard's star may have one or more unseen companions with planetary masses, but this result needs to be confirmed. The errors in the data are too great for the observed systematics to be definitely considered real.

Possibility That Other Stars Have Planets

The suggestions in the literature that other stars might have observed extrasolar planets are not as firmly based as the work on Barnard's star—which, in fact, is a good star to study for extrasolar planets. To see this, let us consider a more fundamental question.

Are all types of stars equally likely to be good candidates to be searched for planetary companions? If we have a group of systems, each with a star and a Jovian-mass planet, the planets that will be most easily detected by the astrometric method will be those with the least massive primary stars. These systems will have the largest ratios of planetary to stellar masses. This suggests that the astrometric method works best for low-mass stars. Such stars have to be fairly close to be sufficiently bright to study. This helps in the detection of their extrasolar planets.

For a triple star system, there are no stable orbits if the distances among the three stars are comparable. For stability to occur, two of them must form a close pair and the third has to be at a relatively large distance. As a planet is a substellar mass, it can be substituted for one star in this arrangement. Consequently, a planet can be in a stable orbit if it circles either star in a widely separated binary system or if it circles both stars in a close binary system. There is no guarantee that planets, if they exist, lie in the zone of habitability for carbon-based life. A further difficulty is that the formation of a planet may not have been possible even in the region of a stable orbit. Thus, Dr. Thomas A. Heppenheimer of the Center for Space Science in 1974[12] and in 1978[13] expressed doubt about the existence of planets in many binary systems where the two stars are separated by less than the radius of our own solar system. This is particularly true if the orbits are rather eccentric.

In the process of star formation, the protostar collapses from a huge, very diffuse gas cloud whose initial diameter may be a light-year or so. The protostar is likely to have some rotational motion. As it collapses, it speeds up, just as an ice skater does when he brings his arms in toward his body to spin faster. Rapid rotation will eventually make the protostar unstable. It can throw off some matter and partly alleviate the problem, but it cannot solve it entirely in this manner. Thus, it appears that the protostar splits into two parts that are most likely about equal in size. Most of the rotational motion is now in the form of orbital motion, instead of axial rotation. This mechanism may explain closely separated pairs of binary stars.

Another type of binary is the widely separated pair. Such systems may be the end product of two protostars that contracted separately but became gravitationally associated before they left the star cluster of their birth.

Dr. Helmut A. Abt[14,15] and his associate, Saul G. Levy, both of Kitt Peak National Observatory, studied 123 solar-type stars to see how many had binary companions. They found that 83 stars had close secondary companions of at least $1/160.01 M_\odot$. There are difficulties in how one should properly account for those less massive secondaries that cannot be directly observed. A possible extrapolation is that all stars have companions in accord with the theory of star formation, as cited.

In 1970 Dr. Stephen H. Dole[16] of the Rand Corporation published the results of a computer study of the hypothesis that the planets were formed from the particles in the protostellar nebula, the cloud of dust and gas that surrounds a newly formed star. He found that his simplified computer code was able to produce solar systems like our own, assuming certain initial conditions. His code was later more thoroughly tested by Drs. Richard Isaacman and Carl Sagan of Cornell University.[17] They considered the formation of planetesimals, as discussed by Dr. Peter Goldreich of the California Institute of Technology and Dr. William Ward of Harvard University.[18]

Initially the dust grains are micron-sized, like those in interstellar space. In the protostellar nebula they produce solid objects a few kilometers across. These bodies accrete additional material from the nebula, but they are also subject to destructive collisions with one another. In these collisions all fragments less than about a centimeter are lost to the region of growing planets, because of the protostar's stellar wind or radiation pressure or both. If there is not enough dust, the objects' growth will be limited. Collisions will dominate, so that the largest objects will be no bigger than a kilometer or so. If there is enough dust, the objects will grow to reach planetary proportions. In a system with a great deal of dust, the planets will sweep up all the material and there will be relatively few small bodies by comparison with the solar system. If there is not quite as much material, but enough to support planet growth, there will be both planets and many smaller bodies as in the solar system.

Isaacman and Sagan found that the results produced depended also on how the density was distributed in the protostellar nebula. Systems similar to our solar system were produced in only a small fraction of the possible cases. Usually, the computer code predicted multiple star systems or planetary systems with larger Jovian planets than in our own or none at all. What is important is that they were able to generate planetary systems for a wide variety of cases including single stars with asteroidal bodies, systems with Jovian planets far from the central star, and multiple star systems with planets. The most massive terrestrial type planets contained five Earth masses. The typical planetary system had about ten members.

From these studies we can conclude that most stars probably have planetary companions, although many may be just asteroidal-sized bodies. There is also a fair chance that other stars have Jovian-sized planets, which will be far easier to detect than the small, rocky terrestrial variety. The task is to detect these extrasolar planets, since astrophysical theory tells us that they should exist.

Proposed Methods For Detecting Extrasolar Planets

At a distance of ten light-years, Earthlike and Jovian planets subtend about 3×10^{-5} and 3×10^{-4} arc seconds, respectively. To actually resolve them would require telescopes with effective diameters of

32,000 meters and 3200 meters, respectively (Matloff and Fennelly, 1976[19]), located in space. These might take the form, for example, of an interferometer system located on the Moon. In this chapter we will consider methods of detection rather than resolution, since before we try to resolve details on an extrasolar planet, we want to know that it exists.

If our line of sight were in the orbital plane of the planet about its parent star, we might be able to observe the transit of the planet over the star's surface. For Jupiter and the Sun, this amounts to an eclipse about 0.01 magnitude deep, in which the star's light would be diminished by 1 percent. It would last for about one day every 12-year period of revolution. The required accuracy can be attained with current equipment, provided one is sufficiently careful. But the eclipse size is so close to the limit of accuracy that it would require independent confirmation.

Frank Rosenblatt of Cornell University in 1971[20] proposed using two-color photometry and three wide-field telescopes for the task. Each telescope would be equipped with an image dissector tube. The stellar magnitude would be measured in two different wavelength regions. A central computer would command each telescope to survey its own group of stars until one of them had a preliminary detection, in which case all the telescopes would be used together as a coincidence detecting system. This system would also tell us something about the distribution of light across stellar discs, but it might possibly be confused by such things as starspots and varying luminosity of the star. Both of the latter, fortunately, have colorimetric responses different from that of a true transit. Other sources of noise include the Earth's atmosphere and those inherent in the instrument. This is a feasible yet very time-consuming approach. If 9000 stars per night were observed, then perhaps one planet per year would be detected. Multiple transits would be required for confirmation.

Edward Argyle of Dominion Radio Astrophysical Observatory[21] in 1974 noted that one might be able to detect light echoes of reflected light from planets due to stellar flares. This method would work for Jovian-type planets in orbit about flare stars that are late M-type dwarfs or for regular light variables with large amplitudes and rapid variations, but not for Jovian-type planets about solar-type stars.

In 1974, Fennelly and Matloff[22] speculated on the possibility of detecting Jovian-type extrasolar planets with radio telescopes, by use of the technique of very long baseline radio interferometry (VLBI). To obtain the required angular resolution, two large radio telescopes must be used—one, say, located on the East Coast of the United States and one on the West Coast. Atomic clocks are used to exactly time the observations. There are problems in combining the signals to make detection possible, but, in principle, they can be solved. If they are, and if one works at 3 or at 15 centimeters, then a Jovian-type planet could be found up to 10 parsecs from the Sun. But atmospheric turbulence begins to affect radio waves in the short wavelength part of the radio window so that this technique cannot be used at much less than 10 centimeters. By going into space one is freed from such restrictions and obtains as well a large possible increase in the baseline between the elements of the interferometer.

We have noted that speckle interferometry is a means of correcting the observed stellar image for the bad effects of seeing. A large number of short-exposure photographs of the star are taken and then a computer processes them to eliminate atmospheric effects. A major limitation of speckle interferometry is that the field of view is quite small—about 3 arc seconds across for the camera on the 4-meter (158-inch) telescope at Kitt Peak National Observatory.

Despite this limited field of view, Dr. Harold McAlister, then at Kitt Peak,[23] noted in 1977 that much progress might be made in the study of extrasolar planet detection by speckle interferometry. In order to directly detect objects with this 4-meter telescope, they must be separated by 0.02 arc-second. Also, Kitt Peak's photographic speckle cameras are limited to a magnitude range of ten, while a range of 20 magnitudes is needed for direct imaging.

Speckle interferometry can furnish observations of moderately separated binary stars with extreme precision. A single speckle observation can measure the relative separation of two binary stars with accuracy comparable to a whole year of conventional astrometric observations. By multiple speckle exposures, it is possible to increase the precision over that for a single one by at least a factor of 15. There are about 100 known binary stars within 25 pc of the Sun whose separation could be measured by the speckle technique. Their orbits could be greatly improved, and possible low-mass, faint companions could be found. Because of image-blending effects, standard astrometic programs could not study these systems.

If, by chance, a distant background star were located in the field of a nearby star, extremely precise relative parallaxes and proper motions could be measured. About 120 nearby stars could be so measured. If the proper motion were large, then a given background star could serve for only a limited time—seven months, for example, in the case of Barnard's star. Thus, speckle interferometry, while it uses a very limited field of view at present, could play a very important role in the detection of extrasolar planets.

The late Dr. Su-shu Huang of Northwestern University[24] calculated that the Sun's motion about the center of mass of the solar system, due mainly to Jupiter, causes radial velocity differences of up to 25 meters (82 feet) per second. This effect would be observable only under favorable circumstances, namely when the line of sight is close to the orbital plane of the planet about its parent star. Unfortunately, an improvement in the measurement of the spectral lines would not necessarily mean a greater chance of detection, as the spectral lines are broadened by turbulent motions and other physical causes far more than the 25 meters per second due to the planet.

Although conventional spectroscopic techniques cannot approach the necessary accuracy, Fourier-transform spectroscopy and polarimetric wavelength calibration can do so in principle. Both techniques provide a wavelength calibration for the smallest wavelength regions that can be resolved in the spectrum. To a large extent, this overcomes Huang's objection for normal, well-behaved stars. The exposure times are long enough to average very short-term, low-amplitude, radial oscillations. Fourier-transform spectroscopy involves, at present, working in the infrared, rather than at optical wavelengths, and is the analog of frequency-modulation (FM) radio. Polarimetric calibration is the analog of phase modulation.

Dr. K. Serkowski of the Lunar and Planetary Laboratory, University of Arizona, in 1976[25] designed a polarimetric radial velocity meter to work on a 1.5-meter telescope and to give an accuracy of 10 meters per second (32 feet per second) in a half hour for a fifth-magnitude star. Four observations of the star give the required accuracy of 5 meters (16 feet) per second. If bright solar-type stars are observed, then Jovian-type planets should be detected.

Dr. Serkowski and his colleagues have built such an instrument. To minimize errors, they placed the instrument on the floor of the observatory's observing room. Starlight passes from the rear end of the telescope via optical fibers to the entrance aperture of the instrument. Their observing program consists of the G-type stars that Abt and Levy found were single stars. They hope to detect lower-mass companions of these stars. Since their method cannot determine the orbital planes of the substellar-mass companions, their work will yield only statistical information on such objects. However, it will be very helpful in predicting which stars should have Jovian-sized extrasolar planets. The

Figure 14-1. Dr. K. Serkowski with the Fabry-Perot Radial Velocity Spectrometer. Research for this project was sponsored by the National Geographic Society. (Courtesy Lunar and Planetary Laboratory, University of Arizona)

stars that show small radial velocity variations will be good candidates for other astrometric methods of detecting extrasolar planets. A larger telescope and an improved instrument would mean a decrease in the time needed to make an observation and an increase in the volume of space that could be effectively surveyed. The data that this program is gathering will be quite important in obtaining a complete description of extrasolar planetary motions once they are unambiguously discovered by advanced astrometric techniques.

Gatewood and his associates[26] have continued their work on the problems of the astrometric technique. By use of a sufficient number of reference stars and proper reduction of systematic errors, they conclude that existing astrometric telescopes can achieve higher precisions than hitherto attained. With their new level of understanding, they have begun a program to find extrasolar planets with the 76-centimeter Thaw refractor of Allegheny Observatory. After three years of work, they have found suggestions that van Maanen's Star and Tau Ceti have unseen and probably stellar companions. Continued observations with the increase in accuracy they have achieved will clarify the situation.

To extend the survey to a larger sample of stars an improved detector is needed, especially if they want to use the same telescope. To this end they have designed a multichannel astrometric photometer (MAP) based on existing technology, and have performed instrumental tests with a single channel version to demonstrate that this technique will work. MAP satisfies the criteria for an astrometric detector as it has high spatial resolution and quantum efficiency, is free from important systematic and random errors, and allows one to measure the locations of a sufficient number of reference stars. The measurement of relative positions is done by the determinaton of the phase of electronic signals.

A photomultiplier tube is placed behind a Ronchi ruling, which has equally wide transparent and opaque spaces, located in the telescope's focal plane. If the width of the spaces is less than the seeing disk of the star, then a series of maxima and minima in the light curve are observed as the star moves across the field of view. A second star as monitored by a second photomultiplier tube gives a similar output. If the distance between the stars is known to the nearest whole number of line spacings, then the exact separation can be found from the modulated signals. This principle can be extended to a large number of stars. One moves photomultiplier tubes around to observe the target star and a sufficient number of reference stars. MAP should develop a precision of 0$''$004 within 10 minutes of observing. This suggests an annual accuracy of about 0$''$001, which is a factor of 3 improvement over the best precision now attainable. MAP's accuracy will improve somewhat with use as the relative positions of each ruling line become better determined.

Serkowski[27] suggested an imaging interferometer as an advanced astrometric instrument. It would consist of two telescopes, each with a movable flat mirror and two stationary mirrors in a vacuum. One could measure light path differences to a tenth of a wavelength. By use of two interferometers and the rotation of the sky, the two perpendicular components of separation between any two stars could be accurately measured.

Gatewood and his associates[26] demonstrated that this interferometer would be able to achieve greater precision than MAP with the Thaw refractor. However, they advocated an alternative design which would achieve almost the same precision at one-tenth the cost. They proposed an 81-centimeter refractor mounted so that its optical axis is parallel to the Earth's axis of rotation. Star light is reflected into the system via a flat mirror. With all the image-forming optical elements in a temperature-controlled vacuum, this design removes all potential sources of systematic errors. Such a system should be able to study stars down to 17th magnitude and achieve a positional accuracy of a few times 0".0001. Besides the probable detection of extrasolar planets, this system would lead to greatly improved parallaxes.

Detecting extrasolar planets from space has certain advantages over Earth-based observations, because of the absence of the atmosphere. Speckle interferometry is no longer applicable, since there is no atmosphere to distort the image. Our discussion of what can be done with a speckle camera on a large Earth-based telescope carries over to a similar-sized space telescope with the advantage that the field of view is greatly expanded. In 1971 Dr. G. L. Matloff,[28] then of the Van Vleck Observatory, suggested that from space, one might be able to differentiate different types of planets by photometry. From space, there is also the possibility of the direct observation of planets. To see if it is possible, we must consider the telescopic resolution. Dr. Lyman Spitzer[29] of Princeton University in 1962 showed that to detect Jupiter in the Sun's light from a distance of 5 pc would require a telescope 30 meters (1200 inches) in diameter. Jupiter would be about 23 magnitudes fainter than the Sun. This indicates that extrasolar planets will be extremely faint. If one went to the infrared at, say, 25 microns, then a 25-meter (1000-inch) telescope would be required.

Spitzer noted that his colleague, Dr. R. Danielson, suggested the use of a large occulting disk located some distance in front of the telescope to dim the light of the presumed primary star relative to the planet to be detected. If the alignment problem could be solved, this would permit Jupiter to be detected 5 pc distant with a telescope about 2.8 meters (110 inches) in diameter. With a longer occulter-telescope separation and a more complicated occulter, one might even be able to detect Earthlike planets. Such a detection of a much fainter source in a star's bright light would have to be confirmed as a planet by a series of subsequent observations that would show orbital motion.

Drs. A. J. Fennelly, G. L. Matloff, and G. Frye[30] in 1975 considered using the Moon as an occulting disk with space telescopes. They proposed to link two telescopes in lunar orbit as an optical interferometer and found it might be possible to detect planets similar to the Earth and Jupiter about nearby stars. Matloff and Fennelly in

1976[19] suggested that planetary detection might be possible with an orbital occulter and a large space telescope mounted on a stratospheric balloon.

In 1978, Dr. James L. Elliot[31] of Cornell University further considered the idea that the dark limb of the Moon could be used as an occulting edge to reduce the background light from the planet's star in the manner described by Spitzer. With a 2.4-meter (95-inch) telescope located in space, this technique could detect a hypothetical Jupiter-Sun system 10 pc away. A sufficient signal-to-noise ratio would take less than 20 minutes of observing. Elliot describes an orbit that would allow almost two hours observing time for any star. It is one that is further away from the Earth than the Moon.

Dr. C. E. KenKnight[32] of the Lunar and Planetary Laboratory of the University of Arizona in 1977 found that a space telescope about 2 meters (80 inches) in size, equipped with an electro-optical system that now can be developed, could be capable of distinguishing starlike images at a distance of one arc second or more from a fifth-magnitude star, and fainter by a factor of 10^9. The primary problem is due to diffraction of starlight by random irregularities of the mirror surface. He proposed that this obscuring light could be decreased by reflecting the light beam from a deformable mirror under computer control—probably a small mirror a short distance in front of the TV-type detector. The problem can be avoided by apodization, which causes the intensity of the reflected light to fall off gradually from the center to the edge of the mirror. KenKnight's system is suitable to both detecting the major planets of the nearby stars and studying their orbital motions and would give immediate answers.

Dr. Gerard K. O'Neill[33] of Princeton University in 1968 described the construction and capabilities of a 125-meter (4900-inch) telescope located in space. He suggested that the instrument have many separate elements, each of which would cover only a small part of the aperture. By investigating various design geometries, he showed that in order to produce the sharpest image with the least side lobes, one should have many elements uniformly distributed in the shape of a parabolic disk. The diffraction pattern due to the main star has alternate light and dark bands. Jupiter at 10 LY would have an angular size of $0\farcs0003$. The corresponding spatial resolution is 4×10^8 meters (400,000 kilometers or 250,000 miles). This type of telescope could detect a Jovian-sized planet by looking for the planet crossing a dark band, scanning regions of constant distance from the star and looking for small signal changes, or looking for the motion of the star about the center of mass—that is, the astrometric method with a resolution of about 10^{-5} arc seconds. Such a telescope could consist of 200 individual glass segments, each 1 meter (39 inches) in diameter and 0.1 meter (4 inches) thick. Three space shuttle flights would be required to carry them up. It would be expensive to develop, and there would be problems in maintaining optical alignment among the optical components.

Drs. G. L. Matloff and A. J. Fennelly[34] reported in 1973 the optimum wavelengths to use for maximizing the signal-to-noise ratio. For Earthlike planets one should observe in the ultraviolet, while for Venusian and Jovian planets the optimum is 0.5 μm in the infrared, and for Martian planets 0.7 μm. A. T. Lawton and others have noted that as one proceeds into the infrared, the brightness of Jupiter relative to the Sun increases and makes detection easier. This suggests that long-baseline interferometry, preferably in space, similar to that of radioastronomy, might be applicable (see, for example, Martin[35]). One place that such interferometers might be built is on the Moon.

More recently, Dr. Ronald N. Bracewell[36] of Stanford University and Dr. Robert H. MacPhie of the University of Waterloo have considered infrared instrumentation for the detection of extrasolar planets. To work effectively in this region, the telescope has to be in Earth orbit to avoid observing through the infrared radiation of our atmosphere. They discussed two designs. The first is a long-base interferometer of Stull and Clark[37] with elements 40 kilometers (25 miles) apart. They showed that in one mode the power received from the planet can exceed that from the star by a factor of 100. The second design is of a spinning interferometer designed for the Space Shuttle. The signal from the star would be constant, while that from the planet would be variable with a known rate. The small variable signal would be picked out. Problems include using infrared detectors that are not close to optimum performance, the need to cool with liquid helium, the need to shield from the Earth's atmosphere, and the background of zodiacal light. To detect Jupiter at 10 pc with such a system and using current techniques would require observations about a month in duration.

Gatewood and his associates[26] also analyzed what should be the space-located successor to their proposed 81-centimeter telescope with MAP instrumentation. They suggest a simple reflector with a diameter of 2.1 meters in combination with an MAP instrument. In a three-hour period for measurement, this instrument will achieve a precision of 2×10^{-6} seconds of arc. This is the size of the effect that the Earth produces on the Sun as viewed from Alpha Centauri.

Thus, we may choose among a wide variety of methods for detecting extrasolar planets. Prime considerations for the choice of a method include cost, probability of detection, and time for confirmation. Transit methods will require at least three events, so that the time for confirmation would be of the order of 30 years. Conventional astrometry, if error free, requires a somewhat shorter period. Methods based upon extremely accurate radial velocities, positions, and optical detection would take considerably less time, from weeks to a few years. Since it is easier and less costly to work on the Earth, we should first consider such methods, and then consider imaging methods in space that could extend the detection observations.

Many astronomers condescend to astrometry as a dull, uninteresting field, although much of their work ultimately depends upon it. The initial instrumentation for the Space Telescope has been directed toward solving the problem of whether the universe is open or closed. Along the same line, much large-telescope time is used to study galaxies and quasars with very little devoted to astrometry. As large-telescope time is in great demand, shifting priorities for scheduling from galaxies and quasars to extrasolar planets would meet strong opposition. Even getting observing time on the Next Generation Telescope for extrasolar planet detection will have its problems unless there is a major shift in attitude among astronomers. This change could be caused, perhaps, by the detection of an extrasolar planet.

The work of Dr. Serkowski and his colleagues at the University of Arizona should yield a number of candidate stars for the search for extrasolar planets. These stars should be observed with an improved astrometric instrument such as the 76-centimeter Thaw refractor when equipped with an MAP system. Gatewood and his associates may well succeed in the search for the first extrasolar planet with such an instrument. Other programs with existing instruments such as McAlister's proposal to use speckle interferometry should be tried. Once we have some positive indications that extrasolar planets exist, large telescopes dedicated to search for these objects should be funded. Since there has been a rapid advance in instrument design in the last few years and this advance is likely to continue for some time, one approach to the study of extrasolar planets would be to construct a 4-meter (158-inch) telescope of proven design similar to the one at Kitt

Peak National Observatory, or a 5-meter (200-inch) Palomar-type telescope. Such a telescope would be suitable for speckle interferometry and polarimetric radial velocity meters. A complementary approach would be to build an advanced astrometric telescope such as the one advocated by Gatewood and his associates. This has the advantage that it is a single-purpose instrument.

Once extrasolar planets are detected, the case for a direct-imaging telescope in space becomes much stronger. The type of planets detected from the Earth will certainly have some bearing on the way we want to proceed with space-based telescopes to observe extrasolar planets. Perhaps KenKnight's apodized mirror telescope, an interferometer, or a telescope with MAP instrumentation will be built. To promote the work on the detection of extrasolar planets, a Solar Neighborhood Institute should be established. This proposal is further discussed in Chapter 18.

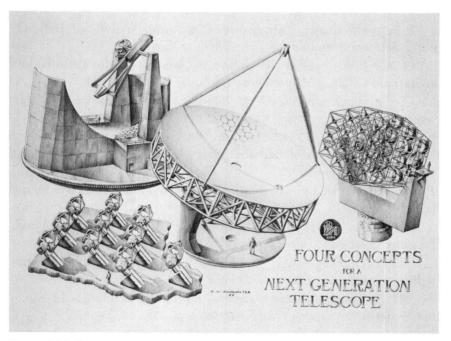

Figure 14-2. Four concepts of next-generation telescopes. These are ideas for new telescopes that will mark a great advance over present ones. The rotating shoe is at upper left, an array of telescopes at lower left, a steerable dish in the center, and a multimirror telescope at the right. (Courtesy of Kitt Peak National Observatory)

Notes

1. S. S. Kumar, "On Planets and Black Dwarfs," *Icarus*, 6 (1967), 136.
2. S. L. Lippencott and J. L. Hershey, "Orbit, Mass Ratio, and Parallax of the Visual Binary Ross 614," *Astronomical Journal*, 77 (1972), 679.
3. P. van de Kamp, "Planetary Companions of Stars," *Vistas in Astronomy*, 2 (1956), 1040.
4. P. van de Kamp, "Astrometric Study of Barnard's Star from Plates Taken with the 24-inch Sproul Refractor," *Astronomical Journal*, 68 (1963), 515.
5. P. van de Kamp, "Parallax, Proper Motion, Acceleration, and Orbital Motion of Barnard's Star," *Astronomical Journal*, 74 (1969a), 238.
6. P. van de Kamp, "Alternative Dynamical Analysis of Barnard's Star," *Astronomical Journal*, 74 (1969b), 757.
7. O. G. Jensen and T. Ulrych, "An Analysis of the Perturbations on Barnard's Star," *Astronomical Journal*, 78 (1973), 1104.
8. D. C. Black and G. C. J. Suffolk, "Concerning the Planetary System of Barnard's Star," *Icarus*, 19 (1973), 353.
9. G. Gatewood and H. Eichhorn, "An Unsuccessful Search for a Planetary Companion of Barnard's Star (BD +4°3561)," *Astronomical Journal*, 78 (1973), 769.
10. P. van de Kamp, "Barnard's Star 1916–1976: A Sexagentennial Report," *Vistas in Astronomy*, 20 (1976), 501–523.
11. G. Gatewood, "On the Astrometric Detection of Neighboring Planetary Systems," *Icarus*, 27 (1976), 1.
12. T. A. Heppenheimer, "Outline of a Theory of Planet Formation in Binary Systems," *Icarus*, 22 (1974), 436.
13. T. A. Heppenheimer, "On the Formation of Planets in Binary Star Systems," *Astronomy and Astrophysics*, 65 (1978), 421.
14. H. A. Abt, "The Companions of Sunlike Stars," *Scientific American*, 236 (1977), 96.
15. H. A. Abt and S. G. Levy, "Multiplicity among Solar-Type Stars," *Astrophysical Journal Supplement*, 30 (1976), 273.
16. S. H. Dole, "Computer Simulation of the Formation of Planetary Systems," *Icarus*, 13 (1970), 494.

17. R. Isaacman and C. Sagan, "Computer Simulations of Planetary Accretion Dynamics: Sensitivity to Initial Conditions," *Icarus*, 31 (1977), 510.

18. P. Goldreich and W. Ward, "The Formation of Planetesimals," *Astrophysical Journal*, 183 (1973), 1051.

19. G. L. Matloff and A. J. Fennelly, "Optical Techniques for the Detection of Extrasolar Planets: A Critical Review," *Journal of the British Interplanetary Society*, 29 (1976), 471.

20. F. Rosenblatt, "A Two-Color Photometric Method for Detection of Extra-Solar Planetary Systems," *Icarus*, 14 (1971), 71.

21. E. Argyle, "On the Observability of Extrasolar Planetary Systems," *Icarus*, 21 (1974), 199.

22. A. J. Fennelly and G. L. Matloff, "Radio Detection of Jupiter-like Extra-Solar Planets," *Journal of the British Interplanetary Society*, 27 (1974), 660.

23. H. A. McAlister, "Speckle Interferometry as a Method for Detecting Nearby Extrasolar Planets," *Icarus*, 30 (1977), 789.

24. S. S. Huang, "Extrasolar Planetary Systems," *Icarus*, 18 (1973), 339.

25. K. Serkowski, "Feasibility of a Search for Planets around Solar-Type Stars with a Polarimetric Radial Velocity Method," *Icarus*, 27 (1976), 13.

26. G. Gatewood, L. Breakiron, R. Goebel, S. Kipp, J. Russell, and J. Stein, "On the Astrometric Detection of Nearby Planetary Systems, II," *Icarus*, 41 (1980), 205.

27. K. Serkowski, as reported in "Project Orion: A Method for Detecting Extrasolar Planets," *Spaceflight*, 19 (1977), 90.

28. G. L. Matloff, "Color Indices of Extrasolar Planets," *Icarus*, 15 (1971), 341.

29. L. Spitzer, "The Beginnings and Future of Space Astronomy," *American Scientist*, 50 (1962), 473.

30. A. J. Fennelly, G. L. Matloff, and G. Frye, "Photometric Detection of Extrasolar Planets Using L.S.T.-Type Telescopes," *Journal of the British Interplanetary Society*, 28 (1975), 399.

31. J. L. Elliot, "Direct Imaging of Extra-solar Planets with Stationary Occultations Viewed by a Space Telescope," *Icarus*, 35 (1978), 156.

32. C. E. KenKnight, "Methods of Detecting Extrasolar Planets, I. Imaging," *Icarus*, 30 (1977), 422.

33. G. K. O'Neill, "A High-Resolution Orbiting Telescope," *Science*, 160 (1968), 843.

34. G. L. Matloff and A. J. Fennelly, "Wavelength Optimization of Signal/Noise for Extrasolar-Planet Detection," *Journal of the British Interplanetary Society*, 63 (1973), 1287.

35. A. R. Martin, "The Detection of Extrasolar Planetary Systems: Part I: Methods of Detection," *Journal of the British Interplanetary Society*, 27 (1974), 643.

36. R. N. Bracewell and R. H. MacPhie, "Searching for Nonsolar Planets," *Icarus*, 38 (1979), 136.

37. M. A. Stull and T. A. Clark, in "NASA Ames Workshop for Extrasolar Planetary Detection," May 20–21, 1976.

15
THE FIRST STARSHIP

In the sixteenth century, Ragusa, a seaport on the Adriatic coast, was famous for the ships built there. The English called them "argosies," a name with romantic overtones. "Starship," too, is a romantic word that evokes visions of voyages to the stars. The first interstellar expedition will be embarking on the most daring adventure in history. News of its progress will be followed by billions of persons. Successful starship travel, however, will depend on solutions to a multitude of unromantic problems in design and construction. Many questions must be settled via the drafting board, the computer terminal, the desk calculator, and the black boxes of the laboratory before the starship can hover in space ready for its vast journey.

Studies on interstellar flight have been conducted for more than two decades by able scientists and engineers in the United States, Europe, and the Soviet Union, who have been fully aware of the trends and potential of contemporary science and technology. The starship's propulsion system, it is reasonable to predict, will evolve from the most satisfactory of the many types that have already been advanced. It is unlikely that the first starship will be powered by some principle unknown to contemporary science.

Choice of Propulsion System

The starship's performance must meet extremely rigorous requirements. It must reach a maximum speed of at least a substantial fraction of the velocity of light. It must be able to cruise at will, independent of its home base in the solar system, have unlimited range, be able to lead an unimpaired existence indefinitely, be a comfortable home for humans for many years, and be able to decelerate to investigate a solar system (or other interesting phenomenon) and then accelerate again to its former cruise speed.

Judged by the foregoing criteria, several types of propulsion can be readily eliminated. A laser-beam spacecraft may do well for missions near the Earth, but a laser-beam starship would be hopelessly "slaved" to its source of power, a giant laser floating in space somewhere in the solar system. The light received would red-shift, becoming steadily weaker as the starship gained speed. Focusing and aiming a light beam over a distance of light-years would not be easy, to put it mildly. If the beam were off target or the ship wandered off it, the ship would be helpless.

The self-contained fusion rocket has a different problem. Because of the limited supply of fuel that it could carry, it could reach only a small fraction of light velocity. The Daedalus probe, for example, would be fueled with 66 million kilograms (30,000 tons) of helium-3 and 44 million kilograms (20,000 tons) of deuterium, yet it could attain only 13 percent of light velocity and would take 50 years to reach Barnard's star, six light-years distant. When it arrived there, it would not slow down but dash through the star's planetary system (if the star has one) in half a day. Project Daedalus was an excellent pioneering study in interstellar flight, but the Daedalus probe is a dead end.

The RAIR, or Ram-Augmented Interstellar Ramjet, a hybrid combining the interstellar ramjet with the self-contained fusion rocket, has been advocated by several engineers. RAIR uses the interstellar gas as thrust mass. In the RAIR, the interstellar gas is electrically accelerated with power supplied by the fusion reactor. The fusion reactor ejects its own reaction products as a separate jet. This is ingenious, but is has the same disadvantage as the self-contained rocket: a limited fuel supply.

The pure interstellar ramjet avoids these limitations by using the interstellar gas as fuel. In a sense, it would sail through the oceans of space. Some scientists have sketched ramjets that would use the interstellar deuterium gas as fuel for the ship's fusion reactor. Unfortunately, deuterium only makes up about 1/8000 of the extremely thin interstellar gas. The task of drawing in enough deuterium to fuel the starship is appalling.

We are left then, with interstellar hydrogen as the starship fuel. Still, the hydrogen nuclear reaction, proton-proton fusion, is fantastically weak, far too weak to power a starship at near-light velocity. The way out of this dilemma may be through a catalytic nuclear reaction, as proposed by Whitmire (see Chapter 12). In this type of reaction, several isotopes act as nuclear catalysts. The heavier nuclei are transformed over and over but never used up. Only the interstellar hydrogen is consumed, and the excess mass not needed to make helium nuclei is transformed into energy for driving the starship. The interstellar ramjet requires solution of other severe problems, such as elimination of drag at near-light velocity, and ionizing the interstellar hydrogen, but they can surely be overcome if the propulsion problem can be solved.

Let us assume, as a working hypothesis, that the first starship is a catalytic nuclear ramjet that evolved from Whitmire's original concept. We will describe here a descendant of Whitmire's proposal as we conceive it after decades of research and development. The interstellar ramjet appears to be the most effective means for powering the starship, and the catalytic nuclear ramjet the most promising type of ramjet. A starship that could efficiently use the interstellar hydrogen could voyage through space as the argosy sailed the oceans of Earth.

Since the main reactor is a sphere 18.4 meters (60 feet) in diameter, the ship's diameter, making allowance for the pipes, conduits, and auxiliary equipment around the big sphere, would reasonably have to be at least 60 meters (200 feet). Here we run into a difficulty. The ship is to accelerate at 1 g for a year to reach a velocity near that of light. If it is to attain its peak velocity, the ship's mass must be held to about a million kilograms (2.2 million pounds). However, to accommodate the fuel tanks, the catalyst tanks, living quarters, supplies, ship systems, and facilities that an expedition lasting for decades will need, the ship must have a huge volume. Its fuselage will carry a huge metallic scoop to draw in interstellar hydrogen, radiators to keep the engine walls cool, grids hundreds of meters across to direct the flow of the incoming hydrogen, and a very large radio antenna.

To keep down the mass to a million kilograms, superstrong composites, far stronger than steel, are required. Composites are compound materials made up of very thin, very strong fibers in a firm matrix. Many composites are, per kilogram (pound), stronger than steel. For example, steel 4340 has a tensile strength of 180,000 pounds per square inch and weighs 490 pounds a cubic foot. Graphite epoxy, a widely used composite, has a tensile strength of 200,000 pounds per square inch yet weighs only 98 pounds per cubic foot. Composites made with graphite have reached a stiffness per kilogram (pound) 11 times that of steel. The advantages of composites are so marked and their future so promising that NASA and the Department of Defense are sponsoring research on new formulations for composites. Fifty more years of research should surely develop composites far superior to even the best known today.

The design engineers would certainly try to eliminate even a kilogram of unneeded mass anywhere in the starship. For example, the main mirror of the ship's telescope might be a concave metallic shell 60 meters (200 feet) in diameter but only a fraction of a centimeter thick, yet as rigid and as finely shaped as the mirror of the 5-meter (200-inch) Hale telescope.

With a diameter of 60 meters (200 feet) to accommodate two reactors, each about 18 meters (60 feet) in diameter, we can estimate reasonably that the ship's length would be several times that—say, 300 meters (1000 feet). Its basic shape would be cylindrical, because of the geometric simplicity of a cylinder and its high volume-to-surface ratio.

The ship's layout would be dominated by the propulsion system. The enormous metallic scoop, a cone over 150 meters (500 feet) across at its mouth, would be mounted on top of the fuselage near the front end. The magnetized scoop would draw in the interstellar gas for over 3200 kilometers (2000 miles) around. The scoop's funnel would lead into a pipe through which the gas would rush into the reactor. There, at a temperature of a billion degrees Kelvin, the interstellar hydrogen would flash though a cycle of transformations at incredible speed, emerging from the glowing rocket nozzle as a blinding white-hot exhaust. The catalyst tanks would be in front of the engine room, so they would help to shield the crew compartments.

Communication over Interstellar Distances

The pilot cabin might be separated from the control center located amidships. The pilot would guide the ship's flight from the pilot cabin while the chief engineer would oversee monitoring the performance of the ship's systems from the control center, where the commander would direct the crew's work. The communications center would be behind the control center. Communications with the solar system would be maintained even when the ship was many light-years away. The ship would have two communications antennas. Outward bound, it would have an antenna mounted on a boom to communicate with Earth and also for radio astronomy. For the return voyage, a larger antenna could be rigged inside the scoop which would then be pointing at the solar system. The antennas would receive and transmit a continual stream of messages.

Communication with the solar system will be sharply different from communication between a solar system spaceship and its base. The solar system spaceship will never experience a delay of more than a few hours between the time it sends a message and the time it receives a reply. At the distance of Pluto from Earth, a spaceship would have to wait 11 hours for an answer to a message to Earth. Since space exploration will engender, in all probability, a network of bases through the solar system, a spaceship would probably never be more than five hours from contact with a base or colony, and most likely much less.

The starship's communication flow will be much slower. At the distance of Alpha Centauri, the crew would have to wait eight years and seven months for a reply to their message to Earth. Just the same, and this contradicts some predictions, they will never be out of touch with the solar system. While the messages will be long in passage, they will get through. The starship and the solar system communication center probably will transmit continually to each other. The messages probably will be pulse coded, each pulse a bit (or binary digit) of information. Despite the cosmic static, the signals will be readily distinguished.

Static may be only a minor nuisance over these gigantic distances. The solar system communication center would probably operate a Cyclops or even several Cyclops-type radio systems (see Chapter 2), so the starship would always be in the beam of one of them. The starship with its 200-meter antenna could receive Cyclop's messages as far away as 800 light-years, much farther than the nearby stars. The starship's antennas should be as large as possible to maximize the flow of information. The antenna might be so designed that it could be expanded and contracted. Expansion would compensate for increasing distance. The ship's transmitter, drawing current from the reactor, might transmit with 20,000 kilowatts. Tens of millions of bits per second could be transmitted by trading bandwidth for power. Even at its destination, the starship could then receive and transmit text, still pictures, and color television. What would be more wonderful than watching astronauts exploring the planets of another solar system?

The critical system of the ship would be its life support–environmental control system. This system would have to be absolutely reliable, not for months, but for decades, as well as being completely self-contained. Unlike solar system spaceships, the starship would not be able to replenish supplies and replace defective equipment from solar system depots. In this it would resemble the Earth itself. While it is obvious that a spaceship is an enclosed system, it is not so obvious that the Earth itself is an almost enclosed system. Humans are not aware that they are living in an enclosed system, for the atmosphere, hydrosphere, and geosphere function so unobtrusively and so effectively that most people take the Earth life support system for granted. People breathe in oxygen and water vapor, consume plants and animals, give off liquid, solid, and gaseous wastes—and Nature takes care of all their needs silently and efficiently. Despite the hazards pointed out by environmentalists, the human race as yet plays a minor role in the Earth's ecology.

The starship's situation will be just the opposite. The ecology of the humans on board will dominate their environment. If fresh air does not flow in continuously, the crew will die of carbon dioxide poisoning. If water is lacking, they will die of thirst; if their food gives out, of starvation. If their wastes are not sterilized, they will eventually die of disease.

Supplying these human needs with a minimum of mass, power,

and volume will be a taxing task even for the science and technology of the twenty-first century. Suppose the starship designers are required to design a life support–environmental control system for a 100-member crew for a voyage to last 20 years. The average crew member will consume per day:

Water, 5000 grams (drinking, in food, and for personal sanitation)
Food, 2000 grams
Air (oxygen), 666 grams
Total—7666 grams per person per day (about 16.9 pounds)

To be on the safe side, let us estimate that the total intake of water, food, and oxygen will be 8 kilograms (17.6 pounds) per person per day. The total required for a 20-year voyage in a starship with a 100-member crew will then be:

Water, 3.6 million kilograms
Food, 1.5 million kilograms
Oxygen, 0.5 million kilograms
Total—5.6 million kilograms (12.3 million pounds)

The starship would, of course, store reserves of food, probably dehydrofrozen and compressed for lightness and compactness; water (which, unfortunately, cannot be compressed or dehydrated!); and oxygen, probably as liquid oxygen. The extreme cold of interstellar space, only a few degrees above absolute zero, would make their preservation very easy. The food compartments would, of course, be well removed from the reactor and well shielded. These reserves would be stored for an emergency. Under ordinary circumstances the crew would rely upon a completely closed life support system in which everything would be recycled. It would be a highly advanced CELSS (Closed Ecology Life Support System; see Chapter 2), by then a mature technology proven by scores of years of experience in solar system spaceships, space stations, and planetary and moon colonies and bases. Still, the starship life support system would be a challenge to the scientists and engineers in this field, for it would have to be potentially eternal.

The CELSS would be partly physicochemical but mostly biological. It would be highly redundant. Oxygen might be supplied and carbon dioxide absorbed by more than 100 species of plants under cultivation. While they would be mostly food plants, grains, vegetables, berries, and dwarf fruit trees, flowering plants would be included. The plant beds would have room for dahlias, tulips, and peonies as well as lettuce, tomatoes, and carrots. The CELSS might center around a small farm, for which the starship's mechanical equipment—such as the pumps, heat exchangers, and conduits—would play a supporting role to the plants and livestock. Probably, the plants would not be confined to the farm. They might be found growing in the oddest places, in boxes or pots, from the control center to the engine room. The livestock would likely be restricted to a few small species such as poultry and, perhaps, dwarf cattle for milk and meat.

The only way to assure the safety of the crew would be to develop a life support system that would function indefinitely. Indeed, all of its systems should last indefinitely. What would happen if the starship could not make the return voyage? This is not impossible for a manned starship. The crew would have to be able to replace any part that broke down. Since the ship might have a score of systems, dozens of subsystems, thousands of pieces of equipment, and millions of parts, this might appear to be impossible.

There are three methods for dealing with this problem: storage, standardization, and duplication. The ship's storerooms could not hold a complete set of replacements, for then the ship would be carrying a complete duplicate of itself as payload! The storerooms would have to hold enough parts to replace predicted breakdowns. Standardization would simplify the problem of replacement by reducing the number of parts to be replaced. Duplication would require that the ship's workshops be so equipped that they would be capable of fabricating any part, system, or machine on board. It would probably have computer-controlled and highly versatile machine tools. The machine tool computers could be programmed to duplicate and assemble thousands of different parts. Besides raw stock, finished and semifinished parts, the storeroom's bins would contain raw materials, metals, alloys, chemicals, plastics, and other items, all salvaged from the ship's discards. All worn parts, shavings from machine tools, cracked plastics, even dust—in short, all the accumulated waste and trash—would be collected and fed into a small fusion torch auxiliary to the reactor. The output of pure elements and compounds would be stored to be drawn upon by the chemical pilot plant for preparation of alloys, ceramics, composites, and chemicals.

Since the starship would have to be a self-contained community dependent upon its own resources yet capable of thoroughly exploring and investigating a solar system, it would have to have the scope of a highly advanced society. It would therefore have research laboratories, an observatory, workshops, storerooms, an information center, a recreation hall, entertainment facilities, a small hospital, and an education center.

Research on Board the Starship

The observatory would be unlike any in the solar system. The main telescope would be deep in the interior of the ship, its giant mirror fixed immovably in the framework. The mirror would extend almost clear across the ship from one wall to the other. Coelostat mirrors in portholes flush with the ship's skin would reflect light through vacuum pipes to the mirror. The observer could view any area of the celestial sphere at will. He would be able to observe, not only in visual light but also across the electromagnetic spectrum from the ultraviolet to the infrared. For gamma rays, X-rays, microwaves, and radio waves, other telescopes would be used.

The information center would be more than a library. Here, in microfiche, microfilm, holograms, motion picture film, slides, bubble memories, and computer tape would be stored the essence of human culture. People would come to look up whatever struck their fancy as well as for research, training or ship operations. The library would contain some books, specially printed on very thin paper, but probably more than 99 percent of its holdings would be in nonbook form. As a concession to human preferences, a printer would read the tape, microfiche, or other form and print it on paper in easily readable form for the reader's convenience.

The amount of knowledge accumulated in one year is monstrous. About 70,000 journals are published each year throughout the world. Let us suppose that a committee of experts chose the most important journals in each field of the life sciences, the physical sciences, the social sciences, and the humanities—two thousand journals in all. Studies by science historians have shown that almost all research papers refer to prior papers less than 20 years old. For selected fields, such as astronomy, the holdings would be more extensive.

Since the average journal prints about 2000 pages a year, 80 million pages would have to be stored in the library.

For standard microfiche, each page is reduced to a twenty-fourth of its original size, so a 10 by 15 centimeter (4 by 6 inch) microfiche card holds 96 pages. Let us suppose that the pages were reduced 2400 times. A wall bookcase, 0.5 meter deep by 2.4 meters high by 17 meters long (1 by 8 by 57 feet) could then hold 80 billion pages, the equivalent of 200 million books, each 400 pages long! This would be more than the combined holdings of the Library of Congress, the British Museum Library, and the Bibliotheque Nationale, in a room no larger than the home library of many of our readers. Besides the sciences, the information center would store the bulk of world literature, art, history, and entertainment. There would be more than enough for every taste, and even decades of voyaging would allow the readers only to sample their fields of interest.

The purpose of the library would not be limited to use of the crew en route. If the crew decided to settle on a planet or moon of a distant solar system or were forced to do so by an accident that crippled their starship, they would have on hand all the expertise they needed, not only of Earth society but also of the societies that would be on the Moon, Mars, and other planets and satellites of our solar system.

The ship would also have a museum collection. All its holdings would be kept in floor-to-ceiling lockers lining the walls. Besides samples of rocks, soils, and minerals, there would be a refrigerated section where the seeds and spores of thousands of species of plants from algae to redwoods would be held. Stored in a vacuum at $-50°C$, they would remain dormant, awaiting the day they would be warmed up and start to grow. There would also be thousands of tiny blocks of plant tissue, their metabolism almost suspended by the deep cold, which could be warmed and grown into duplicates of many food and fiber plants. Besides the plant section, there would be an animal section where thousands of fertilized ova and spores of species from corals up to monkeys would be kept dormant until they might be reared to colonize a sterile planet.

The recreation center would include a gymnasium outfitted for many kinds of games. While the ship was accelerating or decelerating at 1 g, the crew could play any game familiar to us, limited only by space. Even a small swimming pool could be provided. However, when the ship was coasting at zero g, only "space games" could be played. These games would not have to be invented for the starship voyage. The crew would play the games that had been devised years earlier in space stations, where zero g was an ever-present fact of life.

The gymnasium would play a vital role in the crew's daily routine. Everyone would be required to spend an hour a day there when the ship was in zero g, working out under medical supervision. This rule would probably be strictly enforced to make sure that all crew members kept in good physical condition. The gymnasium would have exercise machines for counteracting the harmful effects of zero gravity, subjecting the whole body, especially the muscles and bones, to the stresses they would normally undergo on Earth.

The ship's hospital would probably be long on equipment and short on beds. Those patients who did not require intensive care could be just as well treated in their own quarters. The ship's medical officers could easily make "house calls." The hospital would be furnished with every type of equipment needed for diagnosis, treatment, and operations known to the medical science of the time. It would have to, since it would be the only medical facility the crew would have for the period of their mission. The pharmacy would be carefully stocked with drugs, and a versatile organic chemistry laboratory would be able to synthesize any additional drugs needed.

Amidship would be a facility serving both for therapy and for living quarters. A ring 30 meters (100 feet) in radius would circle the ship's interior. Running smoothly and silently, it would create a partial-g environment for use in the period when the ship's engine was off and the remainder of the ship was in zero g. Each crew member would spend some time each day in the centrifuge. The combination of special exercises and part-time living in a low-g environment, together with diet supplements, would be planned to prevent any deterioration due to zero g even for a period of decades.

The laboratories would be furnished for investigations in fields across the range of the sciences. The scientists would not have to restrict their research to astronomy and the planetary sciences until they reached their destination. They would be able to conduct research on the plants and animals on board as well as conducting experiments in fields such as chemistry and physics. Research might be a popular pastime of the crew.

Studies would not be limited to the sciences. There are vast storehouses of historical data that have not yet been scrutinized. A century and a half of archaeology in the Near East, for instance, has unearthed hundreds of thousands of clay tablets from Ur, Mari, Ugarit, Ebla, Nineveh, and many other sites. Only a small fraction of these cuneiform texts have yet been deciphered. For example, Ebla alone has yielded more than 50,000 tablets and may have many more awaiting the archaeologists. In the next half century, the stacks of unread texts may accumulate until they total millions. Photographs of these tablets would serve perfectly well for analysis by scholars. So, some of the people on the first starship voyage may be happily devoting their off-duty hours to translating into English or another modern language a new-found version of the oldest saga in world literature, the Epic of Gilgamesh. Gilgamesh searched for the secret of immortality and failed. The starship crew, a little wiser, will search for a new world, and succeed.

16
ORGANIZING THE FIRST INTERSTELLAR EXPEDITION

If some reasonable conjectures are made about the first interstellar expedition, many of its problems become apparent. These suppositions are that civilization has reached the level of technology which makes interstellar flight feasible, and that mankind is eager for adventure and possesses the resources needed. The first starship, we may reasonably assume, will not be utterly new, a vessel without any precedents. It will be a descendant of solar system spaceships that have ventured far beyond the orbit of Pluto searching for the Öort comet cloud and investigating the interstellar medium. The engines of these solar system spaceships will be the ancestors of the starship engines.

Choice of Destination

The starship's destination must be chosen before work on its design can be started. By the time civilization is ready for interstellar flight, the search for extrasolar planets may have found many stars with planetary systems. Several of these planetary systems may have been observed for more than a decade. In choosing one of these systems as a destination, the planners will have to consider many factors.

Suppose it is discovered that the nearest star with a solar system is not Alpha Centauri, 4.29 light-years distant, but Delta Pavonis, 19.2 light-years distant? Delta Pavonis has a mass of 0.98 of the Sun's and an absolute magnitude of 4.9, compared to the Sun's 4.4. It is a single star of spectral class G7, while the Sun is G2, so it is a little cooler than the Sun. All in all, Delta Pavonis would not be much different from the Sun as the central star of a solar system.

The decision could be made to head for Alpha Centauri simply because it is the nearest star, planets or no planets. The alternatives, however, might make the decision a hard one. Suppose that Alpha Centauri A has been discovered to have a planet, but it is only about 600 kilometers (320 miles) in diameter, no bigger than a sizable asteroid, while Delta Pavonis has been found to have a system of a dozen planets, two of which are within its ecosphere or life zone. Even more tempting, their diameters and masses are close to the Earth's.

So that we can make the optimum decision when the time comes, we need to maximize our knowledge and capabilities. Specifically, first, the search for extrasolar planets should be pursued vigorously with improved instruments and techniques. A single observation of an extrasolar planet will only establish the possibility of its existence. Repeated observations over a period of years will enable astronomers to determine its orbit, its mass, and other properties. The more techniques and instruments that are perfected, the better, for they will not only serve as checks on each other but will add to our knowledge of extrasolar systems. Second, the higher we can make the maximum velocity of the starship, the better, for maximum velocity converts into the distance that can be covered in a reasonable time. Every increase in maximum velocity will extend the range of the starships and bring more distant stars into the reach of mankind. The discovery that our nearest stellar neighbor, the Alpha Centauri trio, has planets circling each of its three stars would be delightful, but Nature may not serve us so conveniently.

Choice of Building Site

Where will the starship be built? The decision will be made upon the grounds of cost, availability of fuel, and proximity to an industrial complex with its skilled labor force. The British Interplanetary Society study, Project Daedalus, chose an orbit around Callisto, the fourth moon from Jupiter, because the starship was to be fueled with helium-3 dredged from the upper atmosphere of Jupiter. But if the starship does not use helium-3 as a propellant, Callisto has no advantage over sites closer to the Earth. The most satisfactory location could be the lunar space station. The Moon's low gravity and absence of atmosphere, its proximity to the Earth, and its own mines, refineries, and workshops could make it the spaceship construction center for the solar system. Most of the components could be fabricated in the lunar factories and assembled in orbit at the space station together with items brought up from Earth.

The framework would probably be assembled first, attached to the space station for easy access. Much of the work might be done by teleoperators, machines controlled by technicians in the station, with workmen in self-propelled spacesuits doing the more exacting tasks. Once the fuselage was encased in its skin, air would be admitted and the construction crew could work inside in their shirtsleeves. If zero g made space construction difficult, the starship might be built on the Moon.

The Starship Crew

Staffing the ship will be vexing, not for lack of qualified applicants but for exactly the opposite reason. Only a flourishing scientific and technological society can easily develop a starship, and such a society would number its space scientists, engineers, and astronauts by the tens or hundreds of thousands. If recruiting for the NASA Astronaut Corps offers a clue, the starship's recruitment board will have to go through a mountain of applications. Astronaut licenses will probably be as common as airline pilot licenses are today, and many astronauts will have piled up years of experience in space missions. Selecting the best candidates for a 100-member crew from so many competent and eager applicants will be difficult.

It will be much more complex than staffing a solar system spaceship, for the mission will be far more demanding than a mission even as far as Pluto. The duration of the starship's mission places it in a new category of space voyage. Even a flight at half the speed of light to Alpha Centauri would take at least 18 years, allowing minimal periods for acceleration and deceleration and a year's stay in the Alpha Centauri system. Eighteen years is a good share of a lifetime. It is more likely that the first interstellar mission will last about 40 years, about half the average human lifetime today.

Let us assume that the first interstellar mission will last 40 years. The crew, then, can be neither too old nor too young. If, on departure, the average age of the crew is 30, it will be 70 on their return. A 60-year-old scientist might be a fine candidate, but could he maintain top mental and physical efficiency until he was 100? If the average age of the crew were 50 on their return, which would not be excessive (many highly competent astronauts are over 40), they would have to average ten years of age on their departure! How is this conundrum to be resolved?

The requirements can be loosened to permit a solution. While the crew must be highly trained, not every crew member need be at top competence when the starship leaves. They can be further trained over the years. One point is clear: All crew members will have to be highly intelligent and capable of mastering any subject they want to learn, or have to learn. With this in mind, the distribution of ages will have to be such that the average age will not rise too high over the duration of the mission, yet it will always be high enough to ensure a strong nucleus of adult, experienced crew members.

We can conclude that children must be part of the crew on departure. Some children will be born on board, and some will spend half their lives in the starship. The crew will have to run a school that will operate for 40 years and take some of its students from the elementary education level on through graduate or professional school. There will also be continuing education for all.

Time dilation shortens the duration of the mission somewhat. Thus, when the crew returns to Earth, they will find that they are biologically younger than they would have been had they stayed home. The extent to which this occurs depends on the velocity profile of the starship's mission. However, during the mission time dilation has a minimal effect on the crew's daily lives, as they all age at the same rate.

Medical care will center around preventive medicine. Besides treating injuries and sickness, the medical officers will be responsible for keeping the crew members at the peak of physical and mental efficiency, regardless of age. This will require frequent, thorough testing to determine in minute detail the physical status of each individual. With full control of the environment, the medical scientists should be able to completely prevent all infections, whether viral, fungal, or bacterial. By avoiding using known carcinogens in coatings, plastics, devices, food, water, and other possible routes of ingress into the body, and by adequately shielding the ship from cosmic radiation, the designers should be able to reduce the incidence of cancer to below that in contemporary society.

Since pulmonary disease should be readily preventable in a completely controlled atmosphere, keeping the lungs and respiratory tract healthy should not be difficult, and the medical scientists would concentrate on the condition of the heart and blood vessels. Extending the life span in full vigor to over a century may depend upon the state of the individual's cardiovascular system.

The adults are likely to be married, and both husband and wife would be crew members. The children coming on board would be divided equally between male and female. Upon reaching a set age during the mission, possibly 18, they would be expected to marry. At all times, there would have to be a number of women between 18 and 38 years of age. If a habitable planet were discovered and the crew decided to establish a permanent colony, the colonists would have to include women of child-bearing age or face eventual extinction. There would also be the possibility that a wreck or engine breakdown would make a return flight to the solar system impossible.

Since there would be both children and married couples, it would naturally follow that the children would be those of the couples. Otherwise, the couples would not want to go along. Logically, the selection board, which would probably include the chief officers of the starship crew, would select families, not individuals, with the entire family having to meet a complex set of qualifications.

What would be the size of the crew? Science writers have set the number anywhere from four (two men, two women) to thousands, but the range can be restricted to within approximate limits. The size of the crew would logically be determined by the mission plan. Except for very young children, everyone would serve as a crew member. The crew would operate the ship, do research en route and on the way back, and explore the solar system that would be their destination. Even a hundred would be a small number for such a task.

As a minimum, enough crew members would be needed to run the ship efficiently and to do an acceptable job of solar system exploration. The maximum would be the number that could be comfortably accommodated. Consider that tens of thousands of scientists are studying the Earth and its immediate environment—its atmosphere, oceans, rivers, lakes, rocks, mantle, crust, earthquakes, volcanoes, minerals, glaciers, mountains, plate tectonics, and much more. Even a thousand experts would not be excessive for the investigation of an entire solar system that might be larger and more complex than our own.

As spaceships grow in size and complexity, and their missions lengthen, the crew size will accordingly increase. The smallest number that could operate a huge, intricate starship with its many systems and subsystems, and its complete self-dependence, would certainly be greater than that of the first true spaceship, the Space Shuttle, with its three- to seven-person crew. The starship's key systems would have to be monitored around the clock, the life support system would be far more intricate than that of the solar system spaceships, information and education subsystems would be incorporated, and so on.

Some starship analysts would argue for the very smallest crew, others for the maximum possible. The minimalists would contend that the savings in mass could be converted into added equipment and supplies for increased safety and utility, especially for exploring the target solar system. The maximalists could reply that with a totally regenerative life support system, the CELSS, the size of the reserves of air, water, and food would not be controlling. Besides, a large crew would offer a much wider variety of skills, more backups for vital specialists such as medical scientists, and more opportunities for people to get along with each other.

A large crew does offer definite advantages over a small one. Of the 100 fields of academic study listed by the National Science Foundation in its studies of science and engineering in the United States, at least 80 would be useful in operating a starship and exploring a solar system. Experts in all the major fields of science and engineering would be needed, including astrophysicists, astronomers, geologists, mineralogists, biologists, evolutionists, molecular biologists, chemists, microbiologists, meteorologists, physicists, paleontologists, oceanographers, geophysicists, medical scientists, geographers, botanists, physiologists, biophysicists, geneticists, zoologists, anthropologists, archaeologists, toxicologists, pharmacologists, astronautical engineers, chemical engineers, electronic engineers, mechanical engineers, and nuclear engineers.

Even if life were not found in the target planetary system, the specialists in the life sciences would not be superflous. A planet may appear fit for life as observed from the solar system, but the question of whether it is or not can be settled only by a thorough first-hand scientific investigation. If the planet did not shelter life, it might be prebiotic, possessing soil rich in chemicals that could be the forerunners of life on its surface. That would be highly interesting. Or life might have existed on the planet and perished, also an intriguing puzzle. The crew should be prepared for any eventuality.

Skills that could be useful would be so numerous that each crew member might have a primary specialty, a secondary one, and a third as a training assignment. This would be in addition to subjects that he or she might study as a hobby. Certain skills could be required of all crew members, such as paramedical skills so that they could assist the medical officers in an emergency or give aid until a medical officer arrived. They would also be expected to become familiar with the ship's systems and subsystems and be able to detect any breakdown that might be potentially dangerous.

A crew of a few hundred, therefore, would not be unreasonably large. It might have a cadre of the following supervisors:

STARSHIP SUPERVISORY PERSONNEL

Commander
   Deputy Commander
   Communications Officer
   Navigator
Chief Scientist
Medical Director
Chief Engineer
   Propulsion Engineer
   Communications Engineer
   CELSS Supervisor
   Plant Colony Supervisor
   Animal Colony Supervisor
   Food Technologist
   Auxiliary Power System Engineer
   Workshop Manager

Administrative Officer
   Diningroom Manager
   Living Quarters Manager
   Storeroom Manager
Education Director
   Information Director
   Recreation Director

All positions, except perhaps the very top ones, would be rotated to prevent monotony and to broaden backgrounds. A surgeon might work part time as an assistant veterinarian with the Animal Colony Supervisor, while the Assistant Astronomer might help the Communications Engineer and vice versa. Keeping close track of everyone's training assignments would be a responsibility of the Education Director, who would also, among his other duties, keep an updated record of every member's skills, hobbies, supplementary vocations, and interests. He would be responsible for the crew training program. The aim would be to have a backup for every specialty.

Operating a Starship

Most discussions on interstellar flight assume that the flight itself would be dull and that boredom would be a serious threat to the crew's mental condition. This runs counter to experience with manned space flight to date, and here the past may assuredly foretell the future. From Mercury, through Gemini, Apollo, Skylab, and the Space Shuttle, the astronauts have taken on ever more difficult and responsible tasks in operating the spacecraft and detecting and repairing breakdowns. This was one of the most convincing arguments for selling the Space Shuttle to Congress—the ability of astronauts, unmatched by any machine, to retrieve and maintain satellites in space. Moreover, the Apollo and Skylab astronauts complained of being too busy, never of being idle. With far more tasks to perform, the additional inducement of investigation of new worlds and the universe as never before seen by human beings, and the opportunity to follow one's interests at will, no one on the starship would have cause to be bored.

Aside from their routine duties, research would be the main occupation of the crew. The astronomers, including volunteer amateurs, could use the superbly equipped observatory to make observations around the clock of everything visible (or otherwise observable) on the celestial sphere. One continuing project would be mapping the universe, determining the distances to stars, star clusters, galaxies, and dust and gas clouds with unparalleled accuracy. Their space motions, temperature, composition, and structure could also be determined. The medical scientists could use the crew and the animals as subjects of a longitudinal study on the effects of zero g for almost two decades. The engineers would monitor the performance of the ship's systems and try to improve it. The botanists would breed new varieties of plants, and so on.

At one time or another, all adults would teach, and all crew members, young and old alike, would be students. The children would grow up in a culture in which learning was regarded as natural and continuing for life. Besides their formal education, the children would probably be allowed to sit in on any class, no matter how advanced (as long as they were quiet!), even though they might have little or no comprehension of what was being taught. When old enough, each could serve as an apprentice in the specialty, hopefully, of his or her choice.

The starship crew would be a microcosm of society, just as the starship's CELSS would be a microcosm of the Earth's environment. Both would deal on a small scale with the large-scale problems that mankind faces on Earth. Population, education, manpower utilization, and morale are problems of human organization, while energy, pollution control, and recycling are problems of our environment.

A stellar communication network would have to be developed to keep in touch with the starship. Just as NASA expanded its space communication setup to become the Deep Space Network, a Near Stellar Network will have to be devised. It would probably be the outgrowth of a Solar System Network. The Near Stellar Network might not be wholly or even mostly terrestrial. It might be located in space near the Moon, on the Moon, or on other bodies far removed from man-made radio static, in the quietest region of the solar system.

Since the starship would be traveling in a straight line toward a visible target, a star, the Near Stellar Network operators would seem to have an easy job in keeping in touch with it. All they would have to do would be to point their antennas at the star. Unfortunately, all bodies in the solar system both rotate and revolve. Even the Moon, which rotates so slowly that it takes 27 days and 8 hours to turn once on its axis, would soon carry the antennas away from the point on the celestial sphere that marked the location of the starship. But continuing contact with the starship would be necessary. As the starship receded from the solar system, pointing would have to become steadily more precise in order to receive the strongest possible signal from the starship and to make sure that the starship was receiving the strongest possible signals from the Near Stellar Network. Any interruption might make it hard to regain contact with the starship. Confirmation might not be made until a reply was received from the starship. This at first might take days, and later on, years. Also, continuous tracking of the starship's transmissions would give the best possible accuracy in determining the Doppler shift of its signals and thus its velocity.

The Near Stellar Network designer would consider among his or her many choices two in particular—a huge radio antenna floating in space or several antennas located on a body such as the Moon. The latter arrangement would be like that of the Deep Space Network with its three big antennas located in California, Spain, and Australia, so they can always keep a spacecraft in view. A space antenna would have to be controlled on three axes, for it would not be immobile in space. Instead, it would revolve. Depending upon its location, it would revolve around the Earth, around the Moon, or around the

Sun. If it were near the Earth, it would have to be protected from the Earth's radio and television traffic by a huge free-floating reflector shield moving in synchronism with the antenna. The advantages of a space antenna over one located on a solid body in the solar system might be less than is commonly believed.

Still, a lunar antenna would have its own limitations. It could view only a restricted sector of the celestial sphere. Three antennas located on the lunar equator, 120 degrees apart from each other, would be needed to afford a view of the whole celestial sphere. The lunar antennas, however, might be cheaper than the space antenna. If the lunar colony had manufacturing facilities, the cost of transportation to the sites would be less than that for the space antenna.

The best solution might be a combination—an antenna deep in space far from the Earth working with a lunar antenna network. The combination could do what no single antenna could: track the starship with extreme accuracy. By using the technique of very-long-base interferometry, two radio antennas on Earth several thousand kilometers apart can determine the positions of pulsars to within one hundredth of an arc second. The farther apart the antennas, the greater the precision with which they can determine the position of a radio source. An antenna on the Moon, working with another in space at the distance of, say, the orbit of Mars, could track the starship to within one ten-thousandth of an arc second when it was at the distance of Alpha Centauri.

The starship described in this chapter would probably have at least the following systems:

SYSTEMS OF A STARSHIP

Structure	Workshop
Propulsion	Environmental Control
Fuel	Stores
Shielding	Information
Entertainment	Observatory
Exploring Vehicles	Laboratory
Communications	Living Facilities
Central Computer	Medical Center
Auxiliary Power	Recreation

Exploring vehicles would include a space shuttle, rovers, airplane, boats, and small remote-controlled vehicles.

The monetary cost of the starship cannot yet be estimated; too many uncertain factors are involved. Will inflation continue for decades, or will the price level stabilize and then decline under the steady pressure of an energy-rich technology? What will be the cost of energy? Will the world population stop growing while the world gross product continues to increase? Will most of the components of the starship be custom-made or off-the-shelf items? What will be the average price per kilogram of manufactured goods? How much will the starship differ in design from an advanced solar system spaceship? These are some of the considerations that will set its monetary cost.

As science and technology advance, the starship will become feasible. The question "Can we afford it?" will then be asked by those who have no wish to undertake the project. If science has strong support, however, and technology is being intelligently used for the betterment of humankind, when the time comes that we can build a starship, it will be built.

17

THE FIRST INTERSTELLAR EXPEDITION

Once technology has made interstellar flight feasible, it is likely that humans will embark on interstellar voyages. Let us imagine such an expedition. Simply because it is the nearest star, Alpha Centauri is chosen as the target, even though it might not be the best candidate (see Chapter 14) and might not have an extensive planetary system. The system described here is hypothetical, but the number of planets per star is typical of current computer models of solar systems.

The Beginning

The vast circular shield of the Farside Telescope gleamed in the sunlight, a beacon to spaceships millions of kilometers away. On the side away from the Sun, the 30-kilometer mirror floated in its shadow. The mirror was made up of many segments kept in place and finely adjusted by a system of active controls that included lasers. The telescope's

Focal Plane Satellite kept its station precisely at the focus of the vast mirror, 60 kilometers from its center.

The commander of the starship Discovery was not concerned with the mechanics of the satellite's operations. He was standing in the main cabin of the FPS, watching the screen that made up its front wall. A small yellow disk, Alpha Centauri A, glowed in the center of the black screen. Around it were four white dots, the brightest planets of that distant star. The commander nodded, and the technician at the control panel switched the view. A new image appeared: a little bluish-white globe. Tantalizing glimpses of a dark blue sea fringed by green land could be seen through gaps in the white cloud cover. This was Planet Alpha Centauri A-5, the sensational discovery made with the Farside Telescope soon after its completion, five years earlier. The commander looked at it thoughtfully. This planet was the goal of the expedition, the fulfillment of the dream that somewhere, far away, mankind would at last find a fellow Earth.

The starship's crew could, of course, view the planet on their screens, but it would appear much smaller than it did from Farside. The starship's main telescope was huge by mid-twentieth-century standards but a mere toy compared to Farside. The crew had photographs, maps, tables, and compilations of every last bit of data that the astronomers had squeezed out of their instruments. These would have to do until the crew could start making their own improved observations as the starship neared the target star.

The commander turned away. There was much to do before they left. He wished that the elaborate leave-taking ceremonies planned had been canceled. They were not his idea. If he had had his way . . . He shook hands with the staff and walked back to the docking port where the FPS shuttle was waiting for him. Once aboard, he told the pilot to head for Eagle Base, the capital of the lunar colony.

The shuttle landed on the fringe of Eagle Base Spaceport, far from the huge concrete platforms on which the massive spaceliners landed. Four spaceliners were in port, gleaming in the spotlights against the background of the black lunar night. When he came out of the entry tunnel, the commander was not surprised to find a delegation waiting for him. He sighed. Oh well, he thought, soon he and his crew would be back in the starship, which was now attached to the lunar space station.

Twelve hours later, the commander was in the command cabin of the starship, strapped in the pilot seat. He glanced around. On both sides, crew members were at their posts, intently watching their panels.

All of them were strapped into their seats. The cabin was utterly still. The commander turned back and gently eased forward the control lever on his right. They waited. For what seemed an interminable time, nothing appeared to happen. Then the cabin floor began to vibrate gently. On the main screen, the lunar landscape far below began to slide slowly backward. They were on their way.

Billions of watchers on the Earth, the Moon, and throughout the solar system could see a bright light appear in the darkness as the starship's engine began to build up thrust. The plume of white-hot flame shot back two kilometers, spreading out to form a glowing cloud. They could see the cloud slowly diminish in the distance. Soon it would become invisible to the naked eye, then to binoculars, and finally to small telescopes as it disappeared into the void.

Twelve days later, the ship's public address system announced they would soon be passing the aphelion distance of Pluto, the generally accepted limit of the solar system. Pluto was on the viewing screens. Everyone stopped to look. The dark, frozen planet appeared friendly to them. The view switched to Pluto Base, a small island of light and warmth in a wilderness of black peaks and sullen glaciers of frozen gases. The Director of Pluto Base came on screen, surrounded by his staff. He said that the whole staff wished them well and that they would keep in touch with the starship as long as they could.

The screens went dark and the crew silently went back to work. For several hours, no one wanted to talk. They were venturing far out into the deep ocean of space, far beyond all earlier voyages. A brief announcement: the starship had reached the speed of 10,700 km (6650 miles) per second, 3.5 percent of the speed of light.

Six months later, on the 140th day of the flight, the commander called a meeting of the crew in the pilot cabin. Everyone, without exception, was to be there. Those who had duty stations could monitor their posts from the auxiliary panels in the command center. When they had assembled, the commander made a brief speech. At noon the next day the ship would reach its planned velocity of four-tenths the speed of light, 120,000 kilometers (74,000 miles) a second. As soon as that happened, the engine would be shut down, acceleration would end, and the ship would travel at uniform velocity for the next ten and a third years. The transition from 1 g to zero g would be fast, lasting about a day. Everything, and he meant everything, would have to be checked to be sure it was securely fastened down. Each of them had his or her assigned area to check. Any questions?

The starship will be in zero g for 93 percent of the duration of its mission to Alpha Centauri—that is, for ten years and four months. One of the preconditions for interstellar flight will be the proven ability of humans, along with their equipment, plants, and animals, to live in zero g without any ill effects for at least a decade. By the early twenty-first century this requirement should probably have been met, assuming that permanent space stations will have been in existence since the late 1980s and that spaceships permanently stationed in space will have been in service for at least two or three decades.

Let us imagine what life on such a voyage may be like.

A Day En Route

The Watch Officer coming on duty sat down in the pilot's seat, made himself comfortable, and glanced at the time panel. The hour face read 00:00 00 and then 00:00 01 as the day passed from July 2 to July 3, 2035. The Watch Officer settled back in his seat. He would spend the next three hours in the pilot cabin while almost everyone else was asleep, checking the panels to make sure that all the ship's systems were working properly. After a while, tiring of sitting, he left his seat and walked around the cabin, his shoes sticking slightly to the fabric-covered floor. Since it had been agreed that in zero g, the longitudinal axis of the ship would be "horizontal" and the bulkheads "vertical," orientation was easy even though you could float through the air or walk on the ceiling, if you chose. Once mastered, movement in zero g was easy.

As usual, everything was in order. Continual inspection, planned maintenance, checking by the systems computer, and multiple redundance had raised the ship's equipment to a level of reliability that would have surprised a twentieth-century engineer. Engineering was not the main concern of the Watch Officer, although, like all the adults, he was familiar with the ship's critical systems. He was the Assistant Exobiologist, and his overriding passion was to observe the life forms they hoped to find on at least one of the planets of Alpha Centauri A.

The Watch Officer did not have to watch the panels continually. As long as he looked them over every few minutes, he could do as he pleased. The view ahead on the central screen was arresting and very beautiful, but he had seen it for eight years now and the ship was still too far from Alpha Centauri to start close-in observing. For that, they would have to wait several months more. He rummaged around in the drawers under the keyboards until he found a large white pad. He pulled up the folding sidearm of the pilot's seat and started to write.

The three hours passed quickly. Occasionally he turned on a tape of one of his favorite symphonies. His taste in music, like that of many scientists, ran to the severely classical. Every now and then, he put down his pen and walked around the cabin looking for the flashing red light that spelled trouble. He found none.

The Watch Officer was so absorbed in his calculations that he did not notice the cabin door slide open as a young woman entered the room. She tapped him gently on the shoulder. He looked up in surprise. His three hours were up. He barely replied to her greeting, picked up his pad, and went off to catch up on his sleep.

The new Watch Officer was the Assistant Medical Scientist. When not examining the panels, she spent her watch reviewing the crew's medical record. The semiannual medical checkup would start next week, and she first had to review the medical history of each of the crew. The computerized medical file maintained a record of over a thousand medical tests run on each crew member. She decided to start with one of the youngest, a 14-year-old boy. She punched his identification number on the keyboard, and Medical History I.D. 17 appeared on the screen. She scanned it carefully as the test data slowly moved up and off the screen. Nothing wrong so far. She punched the next identification number and scanned the long list of test findings. By six o'clock, when her shift ended, she had reviewed the record of half the crew and had jotted down several pages of notes. By six-thirty, everyone was up and about except for the few who had been on night shift. They were sound asleep while everyone else was having breakfast.

After breakfast the Communications Officer went straight to the Communication Center, where a pile of dispatches that had come in overnight awaited him in the printout bin. He leafed through the text and photographs. News of the Earth, the Moon, and the solar system colonies, a copy of the journal Science, a copy of Nature, the position fixes of the starship sent by the Near Stellar Network, three years old but still useful as a cross-check on their own navigation, personal messages to crew members, a new documentary film on Ganymede, and much more. He leafed through the pile. Strange how unimportant and remote the affairs of the people back on Earth appeared.

He sorted the material for distribution. There was a letter from

WSA headquarters for the Commander, but it would mean little. From the day the ship passed out of the solar system, they had been on their own, freed from Earth control by the enormous distances they would traverse.

The Communications Officer turned to the "out" bin. This, too, was well filled. He arranged the items in the order he wanted them sent and pulled out those he decided could be dispatched later on when traffic was light. Light traffic was infrequent, for the people back home were avidly curious about everything that happened on the starship, no matter how trivial. First to be dispatched would be the navigation data for the last 24 hours, then astronomical observations for the period, which included interesting data on a G-2 star with a preplanetary disk and a gigantic black hole—fortunately 10,000 light-years away. Summary engineering data would follow, then a pro forma report by the Commander and a miscellany of research reports. Nothing unusual. Now if he could only persuade the Commander to let him borrow some more power from the auxiliary power system.

His introspection was interrupted by the sight of a youngster rapidly searching through the "hold" and "out" bins. The 14-year-old, who bore the title of Information Assistant, was the complete staff of "The Discovery Daily Newsletter"—reporter, editor, compositor, production manager, and distributor. He was selecting material, which he would summarize and then run off on the Communication Center copier. After placing the copies he wanted in his own box, he went off on his daily survey to gather the ship's own news.

The boy loved his job, for it gave official sanction to his lively curiosity. Every day he roamed through the ship from one end to the other, looking for interesting items to report in "his" newsletter. He knew every room, corridor, hall, stairway, ladder, and tunnel of the huge ship and was always looking for shortcuts. He decided that today he would start by visiting the Animal Colony. He was in luck. The Animal Colony Supervisor met him with a smile and told him that a calf had been born overnight. The gawky animal, carefully inspected, was only the size of a toy collie, for the ship's cattle were a specially bred space variety. The boy moved down the corridor between the poultry cages, greeted by an incessant din from turkeys, ducks, chickens, and even a few honking geese. Pets had no place in the starship's economy, and the livestock were the children's closest approach to pets. The boy jumped up and glided in midair down the corridor, bathed in a warm breeze. The animal colony had the best ventilation of any part of the ship, and for good reason. He opened the door at the end of the passageway, slipped through, and slid it shut. He was now in his favorite room, the Plant Hall.

The Plant Hall, the largest hall in the ship, was 90 meters (300 feet) long and 30 meters (100 feet) wide with a 10-meter (30-foot) ceiling. Looking down the hall, he could see the green of living plants everywhere—corn, wheat, rye, and oats followed by vegetables in all stages of growth from seedlings to plants ready for harvest, for harvesting was planned to be continuous. Beyond were orchard plots of dwarf fruit trees, subtropical species standing next to temperate ones, grapefruit and oranges keeping company with apples and plums.

The boy walked down the center aisle fascinated. He liked plants. They were alive! He would be a botanist when he grew up, and when the ship reached Alpha Centauri he would be ready to explore the flora of its planets. Unlike some of the adults, he was certain that at least one of its planets would prove to be as fruitful as the Earth.

At the far end of the hall, beyond the farm plots, were the individual gardens. Anyone who wanted one was assigned a garden plot, to cultivate as he or she pleased. The area was a carpet of flowers. Since all the factors that influence the growth and reproduction of plants were under the control of the gardeners, they could indulge their whims as much as they liked. One amateur, remembering his native California, had packed his plot with bushy geraniums, while another had created a small English garden. The young Information Assistant had a singular whim, which touched the ceiling of the Plant Hall. It was an orange tree, the largest plant in the hall. He sniffed appreciatively the fragrance of its white flowers and admired the large oranges set off by the dark green foilage. Orange trees, he believed, were the most beautiful trees in the world. He picked a ripe orange and tossed it from hand to hand. Since these oranges were eaten fresh, they were a variety with a very thin rind. He was tempted to eat it but decided to put it in his pocket for later. He had work to do. If he could only get the Chief Scientist to let him raise a brood of honey bees. They had thousands of bee larvae in hibernation. He would only need a few with a queen bee, but the Chief Scientist had refused, politely but firmly, to break the rule that no animals other than livestock would be reared enroute. So he would have to pollinate by hand every flower on the tree if it were to continue to bear fruit. Not that he minded, but after all!

He was in midair 6 meters (20 feet) above the floor, holding onto a branch with one hand while with the other he transferred pollen from stamens to pistils with a small swab, when the loudspeaker nearby buzzed for attention. The Commander would meet with the crew in the pilot cabin at 16:00. Except for the following people, everyone was to attend. The boy listened carefully. His name was not on the exempt list—not that he expected it to be. Anyway, the meeting should give him some material for the newsletter. Which reminded him that he had better get busy on it, for it was distributed in the evening at dinner.

The Commander opened with a few words and turned the meeting over to the Chief Scientist, who began by briefly commenting on the mission to date. The ship had been running very smoothly, better than expected. No serious systems malfunctions had occurred, and the minor ones were easily resolved. On the whole, performance had been better than nominal. Research, too, had been rewarding. The mapping of the universe within 100 million light-years had been going very well. Extremely accurate data had been accumulating in great volume and were slowly but steadily being transmitted to the solar system as well as being stored for future analysis. The onboard medical, biological, chemical, materials, psychological, and scholarly research was yielding excellent results. Therefore, the first aim of the mission had been largely met.

Consequently, even though Alpha Centauri was still two years away, he suggested that they should start preparing for the exploration of Alpha Centauri A and its system. To begin, a weekly symposium would be held to review all that was known about it to date. Specialists would go over the applicable fields, starting with planetology, then geophysics, geology, geochemistry, meteorology, exobiology, and so on. Everyone should consider the tasks that he or she would like to be assigned in exploring the system and begin to prepare. The Commander had asked the Chief Scientist to start planning its exploration, and he wanted to know at the next meeting what each of them would like to do, for everybody was to participate in the planning as well as the exploration itself.

At the first meeting of the symposium, the Chief Planetologist gave a brief talk on the possibilities of finding fossil remains or even relics of civilized beings on any of the system's planets. "While paleontology and archaeology overlapped on Earth," he remarked, "they might be completely separate in this solar system. Of course, living organisms might be found on Planet Five, organisms that had climbed the ladder of evolution as high as the level of human intelligence—or higher." Doubtless, the explorers would be so fascinated by living organisms that fossils and dead cities, although they might be present, would have to wait. Logically, however, planets bearing evidence of extinct life might be much more frequent than planets with evidence of past civilizations, or contemporary ones. Life on Earth had required over three billion years to evolve to the human level, yet truly human beings had existed for less than three million years or about 0.1 percent of that time. Habitable planets with human-level creatures should be about a thousandth as frequent as habitable planets with lower forms of life, or with no life at all.[1]

The Chief Planetologist warmed to his subject as he continued: "But that is only simple statistics. If human-level Alpha Centaurian creatures have surpassed our level of civilization, they may have settled not only Earthlike planets but also any terrestrial planets, moons, or asteroids. Their relics might even be artificial space colonies. In the latter case, the separation between paleontology and archaeology will be complete.

"Applying this reasoning, we can propose a strategy for exploration. Our first concern should be to search more carefully than ever for outward signs of civilization. As the ship comes within a few thousand AUs, we should step up our monitoring of the electromagnetic spectrum all the way from gamma rays to very long radio waves for anything that could be traffic. The closer we come, the clearer the situation will be, for the lower limit of intensity of signals that we will be able to detect will steadily decrease.

"When we reach the Alpha Centauri A and Alpha Centauri B systems, we should plan to explore them together. This may seem illogical, but it might be sensible to follow a process of elimination. Let us first scrutinize those bodies that cannot harbor life, at least according to our theories. If they prove to be lifeless, they may still show remains of artificial habitations or evidence of past life, now extinct. That, in itself, would be very interesting. If they are lifeless and show no signs of having ever sheltered life or civilization, we will have strong indications that the course of evolution of life on Earth may be applicable to other solar systems, since it requires the same conditions to survive.

"This cannot be proven from a distance. Even orbiting an airless planet or moon in polar orbit at 50 kilometers altitude and mapping its surface with a resolution of 10 meters might not be decisive. Only landings, traverses over the terrain, close examination of rock strata and soils, taking samples and examining them in the laboratory, would enable us to arrive at a definite conclusion. If we do this—and obviously this will take much time, for we will have to land the shuttle at *perhaps twenty sites—then we will be prepared to visit Planets Four and Five.*

"Here, again, we must be careful not to rush to conclusions. Both Planets Four and Five should be mapped with very high resolution before a landing is attempted. The ship should go into orbit far above any trace of atmosphere at, say, 1000 kilometers. Though keeping the ship at geosynchronous altitude would effectively give us the advantage of observing almost a hemisphere from a fixed point, still it might be wise not to let the ship appear stationary if it could be observed by sentient beings. We might be exposing ourselves to danger. By moving relative to the surface, and orienting the ship so that its brightness as viewed from the surface is as small as possible, we minimize the ship's detectability.

"Now let us suppose that no signs of life are discovered even at a resolution of two meters—no straight lines, no curves, no regularities of any kind. The shuttle descends for a low-level reconnaissance and finds nothing that hints of civilization, or of life. The shuttle descends again and this time lands. Extensive ground exploring turns up no evidence of life. We then come back to a paleontological-archaeological search.

"If we should find, let us say, a deserted structure, our first question would be: How old is it? Here, paleontology and archaeology have the same interest. Nature has provided us with both calendars and clocks almost everywhere, if we are clever enough to read them: the natural radioisotopes. Their half-lives range from billions of years to a dozen or even less. In our geochronometry laboratory we can determine in a few minutes the age of a sufficiently large sample to a thousandth of one percent. For example, if the radioisotope we are measuring has a half-life of ten million years, we can determine its age to within a few hundred years under the most favorable circumstances.

"There is another line of investigation that is very exciting. Suppose that in soil samples we find traces of radioisotopes that, as far as we know, have never occurred in a natural state but have been created only by artificial devices such as accelerators or by explosion of fission or fusion bombs—elements such as curium or berkelium. We would then have incontrovertible proof that human-level beings had been on this planet at one time. By radioactive dating, we could tell when they were there.

"We could do even more. By mapping the distribution of these artificial radioisotopes—and here I am talking of extremely tiny concentrations that will strain the sensitivity of our instruments—we will get some insight into the technology of these beings. We might be able to say that this was, for example, the site of a deuterium-tritium reactor. Further, we might be able to hazard an intelligent guess on the history of the site. Suppose that we found traces of a reactor but none of a nuclear explosion. We could speculate that these people were peaceful. But what if we found traces of the aftermath of a 10-megaton nuclear explosion, and no evidence of reactors? This could point to a nuclear war having been fought there. Radioisotopes can be a powerful means for helping us to understand what we shall see—or not see, which may be equally important."

The Chief Planetologist's talk was followed with close attention. Geochronometry would be a popular course.

The following week the CELSS Supervisor spoke. She said she wanted to introduce a proposal for general discussion, one that dealt with planning, not with exploration. They would be traveling back and forth for a year or possibly longer, landing on planets, moons, and asteroids. Would it not be useful to construct a base? As time went on, they would be accumulating huge masses of material that could run into thousands of kilograms. At the base they could be stored, analyzed, photographed, and recorded at leisure. In all probability, the crew would gather more in a year of exploration than the ship could carry for the return voyage. Also, a base might allow a more efficient division of labor. Some people could be out exploring while others were at the base running laboratory studies. With a base, they could have plenty of room for all their activities.

The base, in her opinion, should be built on a moon, not on a planet. The lack of an atmosphere would be a positive advantage, at least in the first stages of exploration. Before the crew could breathe it freely, the atmosphere of a planet such as Planet Five would have to be carefully analyzed, not only by physicochemical techniques that could detect possible toxic constituents to one part in a billion but also by experiments with laboratory animals. Animal colonies would have to live in the atmosphere for a year before the crew could be assured of its safety.

The base, of course, would be provided with an Earth atmosphere. Nor need the base be cramped: it could easily be built to have a square kilometer or more of area with a dome height of 100 meters. There would be space to spare. And there would be other advantages to a base. The crew could rear many species of animals and plants and have them ready for introduction to Planet Five, if that were deemed advis-

able. By keeping them isolated at the base, we could avoid a hasty, unplanned invasion of Earth life that might destroy the planet's ecology, if one existed. And we could build up our food reserves for the return voyage. The engineers could set up an extensive workshop to overhaul the ship's systems and construct new machines and instruments. They could build a very large antenna with a powerful transmitter for communication with the solar system.

The base should be located on a body in the inner Alpha Centauri system, since that would be most accessible to the system as a whole. Also, it should be in Alpha Centauri's ecosphere, which would make an equable temperature easy to maintain.

Lastly, the base could be well stocked and left in a ready-for-occupancy condition. The next starship mission would then be able to take full advantage of the accomplishments of the first. The transmitter could be linked to a network of sensors and a central computer, so it could regularly report back to Earth on the condition of the equipment, the temperature, the atmosphere, and so on.

Detecting a Starship in Space

Since 1947, the American people have been subjected to a propaganda barrage aimed at convincing them that we are being visited by alien beings from "outer space." "Flying saucers" have been "sighted" and photographed while "witnesses" have reported talking to the beings who operate them, always in the witness's native language. These hoaxes and fantasies were explained by the late Dr. Donald Menzel, the distinguished astronomer, and others, as due to mistaking Venus, Jupiter, stars, airplane lights, reflections of the Sun in a distant window, mirages, weather balloons, and so on, in addition to artistic lying, but unfortunately, many people, especially those without a good scientific background, prefer self-deception to the truth.

Let us consider the question in reverse, which will also throw light on the flying saucer myth: Can a starship approaching an inhabited solar system be detected? If this is feasible, the starship's crew might be endangered.

This problem has been analyzed by three British scientists, D. R. J. Viewing, C. J. Horswell, and E. W. Palmer.[2] A starship, or any other object in space, can be detected only by the energy or matter it emits, reflects, or absorbs. Even if the possibility of detecting the engine exhaust or debris from the ship is included, there are no other ways in which a starship can be discovered.

If something thought to be a starship were observed, how would we know it was a starship? In a telescope it would appear as a dot, probably very dim, against the black sky. If the dot seemed to move among the stars, it might be a starship or one of thousands of asteroids. Its velocity could be found by measurements from two stations; we might use the Earth and the Moon, with a base line of about 400,000 kilometers (240,000 miles), or, perhaps better yet, the Earth and a space station many millions of kilometers apart. If the starship were moving at several thousand kilometers a second, its high velocity would immediately distinguish it from the natural bodies of the solar system.

If the suspected starship should suddenly change its speed and direction, that would be significant, for astronomical bodies move smoothly in almost straight lines, except for those in close orbit around other bodies. If the starship had pulsed engines like the proposed Orion ship, its radiation might also be pulsed, but what if it had a smoothly running engine? Or suppose its engine were turned off and it were coasting?

A starship traveling through our solar system at one-tenth the speed of light would have to have a diameter of 10 kilometers (6 miles) to be barely detectable with a 5-meter (200-inch) telescope at a distance of two astronomical units—about 300 million kilometers (186 million miles). If the starship were not transparent, it could be discovered as it occulted a star by passing between the star and the observer. If it were a sphere a kilometer (3300 feet) in diameter, the ship could be picked up in this way only when no farther away than 52 million kilometers (32 million miles), about 0.3 astronomical unit. Three occultations would be the minimum for establishing its trajectory, and a 3-meter (120-inch) telescope would require an hour's exposure for each photograph.

What about radar? The trouble here is that starships will probably be much smaller than even the smaller asteroids. No existing radar could detect a starship entering our solar system. The starship would have to be unequivocally acquired at a distance of at least an astronomical unit to give mankind sufficient warning. The only means of insuring that the starship would be detected would be to observe continually the entire sphere within which the Earth revolves around the Sun. To patrol this vast area, 1000 observing satellites would have to be launched into orbits around the Sun at a distance of two astronomical units. In order to be certain that no starship would pass undetected, each satellite would have to be able to detect any starship that came within a range of 25 million kilometers (16 million miles). If we estimate the cost of these satellites at one hundred million dollars each, the total cost of this detection system, with some allowance for a communication center, would be over $100 billion.

The writers concluded: "No matter how awesome the starship might be in a terrestrial context, it is in its own environment—interstellar space—virtually invisible."

From this, we can safely conclude that if the Alpha Centauri system were inhabited, the inhabitants could not detect an Earth starship even when it was already far within their solar system, unless their civilization were more advanced than ours and had a system in operation for reliably detecting starships. The starship crew would also be able to detect search beams before the starship was positively detected and could therefore take evasive action. There would be no reason for their presence to be known to the inhabitants of another planet if they did not so choose. Thus it is highly improbable that alien starships could be casually observed from the Earth's surface.

Exploring the Alpha Centauri System

In the last stretch of their long voyage, the starship's crew will spend more and more time in observing their destination. Details will appear that no one has ever before seen. Even the most powerful conceivable solar system telescope could not reveal the detail they will discern as the starship nears Alpha Centauri. The case of Mars is instructive. The Martian terrain was clearly seen only when the Viking Landers sent their first photographs.

The first observing phase will be necessarily astronomical—to confirm the findings of the solar system telescopes and to go far beyond them in precision. The distances of the planets from their central stars, the eccentricity and inclination of each orbit, their orbital periods or years, and the other orbital parameters will be firmly determined. The position of each planet, either inside or outside the star's ecosphere, will be apparent.

As the starship approaches, the second or planetological phase will begin once the planets are seen as disks. What will the crew see?

The Alpha Centauri system can be described in some detail. Alpha Centauri is not one, but three stars: Alpha Centauri A, Alpha Centauri B, and Proxima Centauri. The system is 4.29 light-years away; to the naked eye Alpha Centauri A and B appears as the third brightest star in the sky. Alpha Centauri was not known to the ancient or medieval astronomers of the Near East and Europe, since it is too far south. It is a little east of the Southern Cross or, more exactly, 60° 38' South Declination and 14 hours, 38.2 minutes Right Ascension. Alpha Centauri A and Alpha Centauri B are two solar-type stars that revolve around each other in elliptical orbits, taking 80 years for a complete revolution. They revolve around a common center of gravity or barycenter. At their most distant they are 47.2 AU (7 billion kilometers or 4.4 billion miles) apart, while at their closest they are 24.7 AU (3.7 billion kilometers or 2.3 billion miles) apart. The latter distance is about half the distance between Uranus at its farthest from and Neptune at its nearest to the Sun.

Alpha Centauri A closely resembles the Sun. It is equally bright, 1.09 times as massive, and somewhat larger, with a diameter of 1,711,000 kilometers (1,063,000 miles) compared to the Sun's 1,393,000 kilometers (864,000 miles); their surface temperatures are about the same. Alpha Centauri B is smaller than A. It has a diameter of 1,240,000 kilometers (771,000 miles), a mass about 88 percent and a brightness 28 percent of the Sun's. Its surface temperature is 4610K, while the Sun's is 5760K. These figures indicate that Alpha Centauri A is more likely to have a solar system resembling our own than Alpha Centauri B.

Proxima Centauri (Alpha Centauri C) is a much smaller and dimmer star that orbits A and B at a distance of about 10,000 AU. It is a flare star and is not believed to be a good candidate for possessing habitable planets. On the approach, the flares of Proxima Centauri would be carefully monitored. If they proved to be dangerous, and if Proxima Centauri were found to have planets, the exploration of its planetary system would be done by unmanned spacecraft.

Let us assume it has been discovered that Alpha Centauri has ten planets while Alpha Centauri B has only asteroids. If the distribution of planets were different, the expedition would follow a different course. As the starship approaches Alpha Centauri A, the crew decide to view each planet closely and then determine which should be explored throughly.

As the starship swung around Alpha Centauri A, it pointed straight at the fiery globe, presenting minimim surface to its searing heat. The ship was speeding to periastron, its closest point to Alpha Centauri A, when it would be only 16 million kilometers (10 million miles) away. The humans watched the spectacle in the pilot cabin, uneasy and fascinated. By their own calculations they had nothing to fear, but the sight of the glaring star on seven screens, each on a different wavelength, seething and erupting, was frightening. They watched as fountains of fire spurted from a heaving yellow sea of gas, shooting hundreds of thousands of kilometers into black space only to fall back in graceful curves under the grip of strong magnetic fields. Once they saw a white cloud of gas, larger than the Earth, break off and sail majestically into the void. Fortunately, it did not come their way. They were relieved when periastron passed without incident. In a few hours, as the ship hurtled outward, the images of Alpha Centauri A on the screens started to dwindle.

A day later, the starship was 30 million kilometers from Alpha Centauri A and heading for a flyby of Planet ACA-1. A half moon lay straight ahead, expanding by the minute as the ship sped toward it. The image of the planet moved slowly to the left as the starship curved around it. ACA-1 proved to be heavily cratered, its surface a pattern of stark blacks and whites. It was sharply silhouetted against the black sky with not even a faint glow around its disk to show that it had an atmosphere. Despite its 6600-kilometer diameter, ACA-1 was more moon than planet. The blasting heat from Alpha Centauri A had swept away the faintest traces of gas. Everyone agreed they would postpone a closer look till later.

Even so, they were deeply excited. Looking at the screens and through the portholes, they were seeing a new solar system. They had begun a new era in history. Would they find life in it? Intelligent beings? Civilization? ACA-1 was a rocky waste with no sign of artifice. As the ship made for ACA-2, heated arguments broke out. Most of them hoped to find life and even intelligent beings, but the minority maintained that the chances of finding life were almost nil, even on ACA-5.

Two days later the starship was moving in polar orbit around ACA-2, which was smaller than ACA-1 and farther from the central star but just as desolate. It looked like a moon, 5000 kilometers (3100 miles) across, with its rim sharply contrasting with the dark sky. Giant rifts ran irregularly across its rough landscape, splitting craters and even plateaus in two. The planet was a chaos of craters, talus slopes, canyons, and sharp peaks. The crew agreed to put ACA-2 far down on the exploration agenda.

The next morning the Navigator addressed them. Using animated *diagrams,* he demonstrated to the attentive audience that the planets were in such positions that the best course would be to head directly for ACA-10, the outermost planet, and then turn back, visiting or orbiting the other planets as the ship moved back toward Alpha Centauri A. They had only one year to explore the Alpha Centauri A system, and any time that they spent traveling between the planets would have to be deducted from this time block.

A lively argument ensued. Why not explore ACA-5 first? Why waste precious time on moons and moonlike planets? Had they come so far only to view another broiling Mercury, another frozen Pluto? The dispute was becoming acrimonious when the Chief Astronomer intervened. He reminded them that the purpose of the mission was to discover as much as possible about the whole Alpha Centauri system. They had an incomparable opportunity to investigate a companion solar system that would enable mankind to understand its own much more deeply. What had humans known about the planets when they knew only the Earth? What had humans known of the solar system when that was the only one they knew? Were they scientists or were they not?

The Commander took the floor. He stressed how little they knew of these worlds, how coarse and uncertain was their knowledge. It would be wiser, he argued, to examine all the planets before landing on one, not only for their natural properties but also for signs of contrivance, of civilization. They should be prudent, for they had no right to believe that any society they might meet would be friendly. Certainly, none of them wanted to harm any living thing, but they had to assure their own safety. It would be sensible to employ only passive sensing until they were sure that no intelligent civilized beings lived in any part of this system. If such creatures were found, then the crew could decide what to do next without having compromised their security.

The Commander's arguments carried the day. During their long voyage through unchanging interstellar space, the crew's sense of danger had disappeared. Many were shocked by the realization that they might soon be facing unknown risks. Even if they did not find hostile life forms, they might encounter poisonous atmospheres, X-ray fluxes, landslides, volcanic eruptions. When the ship reached ACA-10, two weeks later, the crew were in a very sober mood. Alpha Centauri A was once more only a bright star in the darkness. Looking down, they could barely make out the jagged peaks of the planet in the dim starlight. Close inspection had revealed no signs of civilization, no lights, certainly no structure larger than their resolution of 10 meters.

The starship made its way back to Alpha Centauri A, circling in *turn each of the four giant gaseous planets. Each had a wide belt of rings and more than two moons. Each moon had to be patiently inspected, but no evidence of buildings was found on any of them.*

At last the ship reached ACA-5. It decelerated into an equatorial geosynchronous orbit at 32,000 kilometers (20,000 miles), moving with the globe so it would seem to be a fixed point in its sky, if it could be seen from the surface. The crew had decided to take that slight risk. After a week of exhaustive observing, the starship moved to the opposite hemisphere and repeated its observations. Then it was maneuvered into a polar orbit. When the mapping of the planet was completed, the perigee was lowered to 1600 kilometers (1000 miles) and the orbit circularized. For a week the starship sped around the planet, the crew intently listening and watching, but no signals were picked up by its receivers, nothing but natural static. After a prolonged conference, they agreed to send the shuttle with a crew of ten to fly around the planet, first at 100,000 meters and then at 10,000 meters. After several landing sites had been designated, the safest one would be selected and a landing attempted. The shuttle would be launched the next day.

Notes

1. J. Armitage, "The Prospect of Astro-Palaeontology," *Journal of the British Interplanetary Society*, 30 (1977), 466–469.
2. D. R. J. Viewing, C. J. Horswell, and E. W. Palmer, "Detection of Starships," *Journal of the British Interplanetary Society*, 30 (1977), 99–104.

A good reference on flying saucers is Philip Klass, *UFOs Explained* (New York: Random House, 1976).

18
A PROGRAM FOR SPACE

A most surprising aspect of the American space program is its short-term basis. The Space Transportation System is a spaceline, not a research and development program. The United States has not had a long-range space program since the end of the Apollo Project in 1972. Despite the long lead time required to develop space vehicles before they go into operation (the Space Shuttle began in 1969 and was planned to start full operation in 1982), the intricate planning that enlists thousands of companies, scores of universities, and tens of thousands of scientists, engineers, technicians, and skilled workmen, NASA leads a hand-to-mouth existence at the mercy of Congress and the Administration. The agency can count only on the funds voted it for the current fiscal year.

American opportunities in space are huge—and yet extremely demanding. If this nation were now to embark on 50 years of vigorous activity, this entire effort would suffice only to pioneer in the exploration and settlement of the solar system and to launch the first starship, and even then only if backed by a joint effort of the Free World's industrialized nations. Today, the United States is far behind where it could be in space, because of its concentration on narrow missions and its neglect of the basic problems of space mastery.

NASA Space Planning

Ever since its establishment on October 1, 1958, NASA has been contending with the complex issues of space venturing. From the start, NASA programs were the product of close cooperation between NASA, universities, and industry. The NASA-university-industry team had to learn first how to launch rockets into orbit. These rockets bore little satellites that Russia's Premier Khruschev derisively called "grapefruit." He did not long have cause for derision. President Kennedy's announcement of the Apollo Project on May 25, 1961, focused the team's efforts on a sharply defined goal with a firm deadline, and that concentration of effort paid off handsomely. The first Moon landing came six months before the December 31, 1969, deadline set by President Kennedy. The United States surged ahead in space and gained worldwide prestige. Logically, NASA's appropriations should have been increased—but, for NASA, nothing fails like success. Instead of advancing in space, America retreated. The Apollo Project was terminated. Once manned space exploration ended, public interest in space dwindled. Only with the advent of the Space Shuttle and its promise of many more people engaged in space flight have the American people renewed their interest in space.

The stop-and-go nature of NASA's operations has seriously hindered American space progress. NASA managers worry about the breakup of teams of scientists and engineers who have worked together for years. After a project is completed, there is no assurance that the team will be assigned a successor. The lack of secure employment and the recession in most physical science fields since about 1970 has weakened America's most important resource, scientific manpower. Not only have students failed to enter these fields, but many Ph.D.s have been unable to find suitable employment.

How should we plan our efforts in space? For more than two decades this question has been addressed by NASA officials, Congressional committees, aerospace engineers and scientists, the National Academy of Sciences, aerospace firms, and scientific and engineering societies. Most of these studies have not looked far ahead. Conclusions have been drawn from timid extrapolations of "present technology" and show an astonishing lack of confidence in science and technology.

Not one American astronaut went into space from the close of the Apollo-Soyuz encounter in July 1975 until the spring of 1981. Morale sagged and many astronauts resigned. Not only did the astronauts lack practice, but so did thousands of skilled workers engaged in launching, tracking, communicating with, and servicing manned spacecraft. Meanwhile the Soviet Union launched nine manned spacecraft, sent up 18 cosmonauts, and broke the world space endurance record with a stay of 175 days in space. Their space team of thousands of scientists, engineers, technicians, and skilled workers kept working and became more proficient. Their Salyut-6 has become a space station which is supplied regularly by their new unmanned Progress tanker. In addition the Soviet Union has started development of a space shuttle which may be in operation by 1985.

This was not the only serious interruption of NASA programs. The nuclear rocket project, ROVER, began in 1955. When it was stopped in January 1973, it was not far from ready for a demonstration flight. According to Harold B. Finger,[1] who was the Manager of the Space Nuclear Propulsion Office, while NERVA was not ready for a flight test in 1973, the NERVA engine had been proven. The big issue was the duration of the fuel elements when the engine ran at top power and specific impulse. This goal was achieved to the satisfaction of the engineers. The remaining problem of designing a nuclear rocket stage was not severe. A large hydrogen tank covered with sufficient light insulation had to be built.

In the early 1960s when NASA began to work on manned space flight, studies were begun on a biological life support system that would be vital for manned space stations and long manned missions. In 1965, when the politicians decided that economizing was more important than America's future, long-range space programs were slashed. From 1965 to 1979, almost nothing was done on this vital space system.

Annual authorization and appropriation, year-by-year planning and execution, may do for routine, nonscience agencies but not for dynamic, science-oriented agencies. NASA needs a long-range program that recognizes the constantly changing nature of space research and development. In a framework set by decades, major projects could be realistically authorized and their total costs approved with allowance for possible future inflation. A broad approach to space planning is long overdue. Congress could achieve it by adopting the envelope concept the European Space Agency has employed so successfully.

European Space Agency Planning

ESA's planning procedure differs widely from NASA's.[2] In some ways, planning by ESA is much more troublesome than American space planning, since ESA is made up of 13 sovereign states with two more unofficial member states. And approval of new programs requires the unanimous vote of the Council, on which all the members are represented. Yet ESA planning is much more effective than NASA planning. The reason is "envelope."

When the Council approves a new program, it is approved—period. With the program approval goes acceptance of a total appropriation or "envelope" to pay for it. The ESA Executive Officer assigned the project carries it on with the program envelope. Unlike NASA officials, he is assured that there is little chance the project will be cut down before completion. Each year, the Council votes on and unanimously approves the ESA annual budget. If a cut is made in a program for one year, it is usually made up the next year. ESA has much more freedom of action than any national space agency. If an ESA member country pays only 30 percent of its annual contribution one year and agrees to make up the deficiency the next, ESA can get a short-term bank loan without difficulty. Moreover, the delinquent country has to pay the interest on the loan, not ESA. Further, the envelope provides for inflation. Once a program has been approved by the Council, the member states are committed to carrying it through, inflation included. Since each member country has its own rate of inflation, the ESA accountants have a hard time calculating their contributions, but they do so. Unless ESA overruns the envelope appropriations by 20 percent plus inflation costs, the member states cannot withdraw from the program. To date, no such overrun has occurred.

The ESA envelope would, of course, have to be altered to comply with federal legislation and principles, but there is no reason it could not be adapted to fit the American situation. Planning by the decade or decades instead of the year would keep space activities on a steady level, which would improve the effectiveness of the government-industry-university team. The Space Transportation System shows what can be done along these lines. Transition from irregular launch spectaculars to regular scheduled flights can take space out of the headlines and make it as much accepted by the average person as the airlines. Continuity would enable NASA, industry, industry, and the universities to do long-term manpower planning. Future scientists, engineers, and technicians could be offered secure careers in the space field.

Space enterprise is intrinsically a large-scale enterprise, and skimping on investment always run up the costs in the long run. This applies especially to space propulsion. Francis C. Schwenk of NASA's Office of Aeronautics and Space Technology said in this regard:

The future of space propulsion depends upon what we want to do in space. The nuclear engine's advantages are not as great as they were figured to be, but the advantage lies with high specific impulse which is necessary for large-scale missions. In turn, to justify the development of the nuclear engine, going to the Moon, for example, with large-scale operations would be required. Under this circumstance, nuclear energy would have an advantage for very large expeditions that require lots of material to be carried to the Moon. The savings in cost would have to underwrite the investment.

Using nuclear energy for the Mars mission would require investing a large sum early in the program to provide the propulsion systems in a timely manner.[3]

In brief, the American space program should be (1) *firmly founded*—centering on the big, basic problems, not the secondary ones; (2) *long lasting*, since the really great problems will take years to solve; (3) *large scale*, since space requires the efforts of hundreds of thousands of people, not only in the United States but throughout the western world; (4) *profitable*, so the benefits will exceed the direct costs; and (5) *intriguing*, to capture the enthusiasm of the whole population. Let us see how these criteria apply to several proposed space programs.

Michaud's Space Plan

Michael A. G. Michaud,[4] a State Department official, has proposed an extensive plan for exploring and colonizing the solar system and for interstellar flight. His schedule, shown in Table 18-1, is intriguing.

Michaud follows the accepted sequence for unmanned planetary exploration: flyby, orbiter, lander, rover, and surface sample return, except that he omits flyby. This is logical, since there will be no reason to send a spacecraft on a flyby mission for a fleeting glimpse of a celestial body if the spacecraft is able to go in orbit around it. We can expect that flybys will be dropped as space technology advances.

The future of unmanned space exploration may not be a straightforward extrapolation of its first two decades, as Michaud suggests. Let us imagine that Congress has authorized NASA to "go all out" on unmanned exploration of the solar system and has voted the funds required. What would be the most efficient way to proceed? Certainly, not by designing each spacecraft as a unique type. Spacecraft are already partly standardized. NASA engineers could carry standardization much further and save money while doing so, since a large number of spacecraft would be built. Orbiters could contain landers, and rovers would be nested in landers. Possibly within a few years NASA could launch a fleet of spacecraft to all the solar system targets in Michaud's schedule from 1980 to 2040, 48 missions in all.

The spacecraft would be launched by the Space Shuttle in low Earth orbit. If the spacecraft had a mass of more than 25,600 kilograms (65,000 pounds) or were over 18.3 meters long by 4.5 meters across (60 feet by 15 feet) folded, it could be designed as two modules. The Shuttle would carry the first module to a parking orbit or to the space station, leave it there, and bring up the second in a later flight. The modules would be fastened together, checked, and sent on their way. The reliability of these missions would be very high.

The cost of each mission could be substantially reduced by launch from an Earth orbit instead of the Earth's surface, and by standardization. It might be objected that standardization would be impractical, since the planets, moons, comets, and asteroids are so different from each other. But they are not so different from the viewpoint of the spacecraft designer. The targets can be classified by the environments expected to be encountered. For example, *dense atmosphere–chemically active* would include Venus, Jupiter, Saturn, Neptune, Uranus, and probably Titan. *Low gravity–no atmosphere* would include all the moons except Titan and possibly one or two others, the larger asteroids, and Mercury. Mercury, the moons, and the asteroids resemble each other in their airlessness, extremes of temperature, low gravity, and solid surfaces. An automated lunar rover, modified for temperature range and gravity level, could climb the slope of a Mercurian crater almost as readily as it could a lunar crater. Truly planetary spacecraft could be developed by standardization.

The intersteller probe is a very different matter. Unless it is to report no farther than about 100 AU from the Sun, it must have nuclear propulsion. With a chemical rocket engine, the scientists would have to wait decades to get their data.

Michaud's manned missions are loosely related. His predictions for the 1980–1985 period are reasonably certain, unless Congress becomes even more antispace than it has been; the latter is unlikely,

TABLE 18-1. Humanizing the Solar System: A Possible Pattern of Missions

Launch Date	Unmanned	Manned
1980–1985	Comet probe (1980) Lunar polar orbiter (1980) Space Tug (I) Venus Orbiter/probes (1983) Mars high-resolution/high-inclination orbiter Venus radar mapper Lunar colony site geochemical lander or surface sample return	Space Shuttle (I) Spacelab Free-flying modules Space station (within Shuttle range)
1985–1990	Venus buoyant probes (1985) Mars colony site surface sample return (1984) or geochemical lander Asteroid rendezvous (1986) Venus geochemical lander Lunar colony site rover Out-of-ecliptic probe Jupiter orbiter (1981)/buoyant atmospheric probes Heavy lift vehicle	Man-rated Space Tug Manned orbiting observatory Orbiting assembly/launch platform Space factory (I)
1990–1995	Mercury orbiter (1987) Venus surface rover Phobos/Deimos sample return (1990) or geochemical lander Jupiter moon (Callisto?) orbiter/lander (1990) Mars colony site rover Asteroid belt surveyor Saturn orbiter (1985)/ring rendezvous/buoyant atmospheric probes	Space Shuttle (II) Lunar-capable space tug Moon base Experimental satellite solar power station
1995–2000	Venus surface sample return Mercury geochemical lander Asteroid belt lander Uranus orbiter/buoyant atmospheric probes Ultraplanetary probe	Space factory-colony (II) Manned Mars landing and return
2000–2005	Mercury surface sample return Jupiter moon surface sample return Saturn moon (Titan?) orbiter/lander Comet nucleus geochemical lander	Geosynchronous satellite solar power stations Lunar farside observatory Man-rated interplanetary tug Manned Venus orbiter

Date	Missions
2005–2010	Asteroid belt sample return Neptune orbiter/buoyant atmospheric probes Mercury base site rover Large asteroid (Ceres?) rover Uranus moon orbiter/lander Oort cloud surveyor
2010–2015	Jupiter moon rover Saturn moon sample return Neptune moon (Triton?) orbiter/lander Comet nucleus sample return
2015–2020	Uranus moon surface sample return Interstellar probe Pluto orbiter/lander
2020–2025	Saturn moon rover Neptune moon surface sample return
2025–2030	Uranus moon rover Pluto surface sample return Oort cloud geochemical lander
2030–2035	Neptune moon rover
2035–2040	Pluto rover
2040–2050	Oort cloud sample return
2050–2060	
After 2060	Colonization of Oort cloud/out-of-ecliptic Manned Interstellar flight

Date	Activities
	Space factory-colony (III) Venus orbiting station/terraforming experiments
	Phobos/Deimos station Self-propelled space colonies Interstellar probe launch/tracking infrastructure
	Translunar self-propelled space colonies Mars colony Venus terraforming begins
	Manned asteroid belt tour/landing Mars atmospheric terraforming begins Man-rated outer solar system tug (asteroid belt and beyond)
	Asteroid colony Mercury base Manned Jupiter moon mission Self-propelled space colonies in asteroid belt
	Asteroid mining Solar system trade
	Jupiter moon colony Manned Saturn moon mission Self-propelled space colonies beyond asteroid belt
	Self-propelled asteroid Saturn moon colony
	Uranus moon colony Neptune moon colony Transport of outer solar system ice into inner solar system

Note: The dates in parentheses are taken from Space Science Board, National Research Council, Scientific Uses of the Space Shuttle (Washington, D.C.: National Academy of Sciences, 1974).

since expanding revenue from the industrial use of shuttles will induce politicians to favor space as a source of jobs and national income. The line of progress that Michaud traces for 1985–1995 is also reasonable, assuming that nuclear propulsion is still blocked, but from 1995 on, his sequence seems to be too hopeful. We cannot travel beyond the Moon with chemical rocket engines unless we are willing to pay for extravagantly expensive space flight, whereas the only way to gain public support for space is to lower the cost of space flight as much as possible.

This argument holds even more strongly for interstellar flight. Michaud envisages stellar exploration as occurring in four stages: astronomical, beginning with a search for extrasolar planets about 1980; interstellar probes, beginning about 2010; manned spaceships, starting about 2040; and colonizing of nearby solar systems, starting about 2070. The first interstellar colonies would start the cycle of exploration-colonization all over again, spreading human civilization through the Galaxy. But it is difficult to conceive space flight leaping in one bound from hydrogen-oxygen engines to advanced fusion engines. Further, Michaud has not taken into account the work already done and now underway in searching for extrasolar planets.

Forward's Space Plan

In July 1975 the Subcommittee on Space Science and Applications of the House Committee on Science and Technology requested the American science community to offer proposals for future programs in space. The boldest plan was presented by Dr. Robert L. Forward, Senior Scientist of Hughes Research Laboratories.[5] Dr. Forward testified at a subcommittee hearing, and his plans were published as a House document. Unfortunately, they received little attention from the media and they were not considered by Congress. Possibly the Congressmen were afraid even to discuss a 50-year program for interstellar exploration. Forward wrote:

> *A national space program for interstellar exploration is proposed. The program envisions the launch of automated interstellar probes to nearby stellar systems around the turn of the century, with manned exploration commencing 25 years later. The program starts with a 15-year period of mission definition studies, automated probe payload definition studies,*

and development efforts on critical technology areas. The funding required during this initial phase of the program would be a few million dollars a year. As the automated probe design is finalized, work on the design and feasibility testing of ultrahigh-velocity propulsion systems would be initiated. Five possibilities for interstellar propulsion systems are discussed that are based on 10- to 30 year projections of present-day technology development programs in controlled nuclear fusion, elementary particle physics, high-power lasers, and thermonuclear explosives. Annual funding for this phase of the program would climb into the multibillion-dollar level to peak around 2000 A.D. with the launch of a number of automated interstellar probes to carry out an initial exploration of the nearest stellar systems. Development of man-rated propulsion systems would continue for 20 years while awaiting the return of the automated probe data. Assuming positive returns from the probes, a manned exploration starship would be launched in 2025 A.D., arriving at Alpha Centauri 10 to 20 years later.

Forward emphasized that the Space Age has come about because of advances in science and technology, not because the United States has a long-range plan to embark into space. Nonetheless, we could have achieved far more in space to date if the American technological effort had not been turned off and on again and again, and if the space program had been maintained on a steady level, concentrating on solving the basic problems of space use rather than on individual missions.

Forward's schedule is more sharply focused than Michaud's:

1975
    Interstellar mission definition studies
    Initiate search for planetary systems around nearby stars
    Feasibility design studies of automated probe payload
1980
    Development of critical probe technologies
    Prototype design studies of automated probe
    Design of interstellar communication system
1985
    Prototype design studies of propulsion system
    Fabrication of prototype automated probe payload
    Fabrication of 1/10 scale prototype propulsion system
1995
    Test prototype by 0.1 (light-year) round-trip search for life in solar system
    Start fabrication of interstellar probe vehicles
    Launch of Alpha Centauri probe
2000
    Launch of probes to Barnard's star, Sirius, and Lalande 21185
    Start 10- to 20-year orbital life test of manned spacecraft design

2005
   Start manned-mission definition studies
   Design man-rated propulsion system
2010
   Fabricate and test man-rated propulsion system
   Prepare communication receiver for probe data return
2015
   First data return from Alpha Centauri probe
   Design manned exploration vehicles
2020
   First data return from Barnard's star probe
   Start fabrication of manned exploration vehicles
   First data returns from Sirius and Lalande 21185
2025
   Launch first interstellar manned exploration expedition

Forward applied engineering analysis to outlining a schedule for the design and construction of manned starships, It went deeper than Michaud's program, since it stressed means, not ends; technology, not missions. Forward assumed that progress in fusion would lead to practical fusion reactors and that the interstellar program should be concerned only with space technology.

Proposed Fifty-year Space Program

We suggest an approach based upon fundamental research. We make no claim to originality. Along with our own ideas, we have incorporated many advanced by Hunter, O'Neill, Michaud, Ehricke, Matloff, Fennelly, Whitmire, Martin, Bond, Forward, and others as well as the NASA studies. Our proposal is for a vigorous space program that will involve all of Western society and tap its only real resource, skilled manpower. In time, all mankind may be led to take part.

    First, a goal that will challenge mankind must be set. We believe that the goal should be a voyage to the nearest star that has a planetary system. Exciting goals such as a landing on the Moon fascinate people throughout the world. The absence of such goals leaves them indifferent. The belief that men may travel to the stars disturbs those who want to live in a comfortably closed little world, while it exhilarates everyone who believes that humanity's greatest achievements lie ahead, not behind. The Apollo missions were terminated because they had succeeded, not because they had failed. NASA was talking of extending the stay of the astronauts on the Moon. This was dangerous, for it could lead to a small lunar base that would have to be supplied. This was too great a commitment to space for most politicians.

The way to the stars leads through the laboratory. Consider the schedule shown in Table 18-2. The schedule is evolutionary, not chronological. The progression is: basic research ⟶ applied research ⟶ space propulsion ⟶ space vehicle ⟶ exploration ⟶ settlement. Every advance in research causes a corresponding advance in space technology. To illustrate: chemical engines make a space station feasible as well as travel in Earth-Moon space. Nuclear rocket engines make the manned exploration of the nearer planets practical. Advanced fusion reactors would power the interstellar ramjet that could take men to the nearer stars.

Dates have been purposely omitted from the table except for the overall limits, 1980–2030, since the proposed program is not time dependent. Many writers have predicted that by such and such a date we will have fusion, solar power satellites, a lunar colony, et cetera. All we have to do is to wait patiently and all these things will come to pass. This is a delusion. Science and technology advance only through human labor. If the United States has a strong space program, as it did in the days of the Apollo Project, we will make fast progress in space technology. If we do not have such a program, we will eventually be outdistanced by the Soviet Union.

The point is that technology is phylogenetic. Every device stems from a predecessor. Our proposed program proceeds through successive generations of vehicles to culminate in the starship. These advances are marked by order-of-magnitude increases in specific impulse, which is a good indicator of space mastery. Moreover, there is no sharp distinction between solar system flight and interstellar flight. We see a distinction because we have just begun to travel in the solar system. The ancestor of the starship will be the solar system spaceship doing 1600 kilometers (1000 miles) a second.

Research would be conducted simultaneously in many fields. Research efforts in chemistry, nuclear physics, and plasma physics are largely independent of each other. Research in these critical areas should be pushed as much as possible, above all by bringing in as many scientists, engineers, technicians, and skilled workers as can be usefully employed in them.

Since NASA is doing all its propulsion research and development on chemical engines, this area is in better shape than plasma physics,

TABLE 18-2. To the Stars by 2030: A Proposal for a Long-Range Space Enterprise Program

Central Research Field	Space Propulsion	Space Vehicle	Space Exploration	Space Settlement	Earth Feedback
Chemistry	H_2O_2 engines Dual expander engine Advanced DEE $I_{sp} \cong 500$ sec Hydrogen technology	Space shuttle Single-stage-to-orbit shuttle Orbital transfer vehicle Lunar spaceship	Unmanned: Closer planets Planetary satellites Comets Asteroids Manned: Moon	Earth-Moon space: Space station Space factory Space base Lunar base Lunar space station	Hypersonic flight Hydrogen technology
Physics Phase 1	Solid-core engine $I_{sp} \cong 800$ sec Gaseous-core engine $I_{sp} \cong 16,700$ sec	Interplanetary spaceship 160 km/sec (100 miles/sec)	Manned: Mars Mercury Venus (from orbit) Asteroids Comets Callisto and outer moons of Jupiter Moons of Saturn	Lunar colony Phobos base Mars base (colony) Venus space station Callisto base Mercury base Asteroid bases (?)	Gas fuel reactor
Phase 2	Fusion engine $I_{sp} \cong 167,000$ sec	Solar system spaceship 1600 km/sec (1000 miles/sec)	Manned: Moons of Uranus Moons of Neptune Pluto and its moon Out to 1000 AU to search for Oort cloud of comets	Saturn satellite base Uranus satellite base Neptune satellite base Pluto base	Fusion reactor Fusion torch
Phase 3	Advanced fusion engine $I_{sp} \cong 7,215,000$ sec Annihilation engine $I_{sp} \cong 27,500,000$ sec	Interstellar ramjet 66,400 km/sec (45,000 miles/sec) Annihilation starship (?) (0.9c) 280,000 km/sec (166,400 miles/sec)	Nearest star with planetary system		Advanced fusion reactor Heavy-ion reactor Annihilation reactor (?)

which is critical to the future of space flight. In the field of hydrogen technology, metallic hydrogen, so far produced in the laboratory in thin films under tremendous pressure, would be far more energy-concentrated than liquid hydrogen, although its cost might exceed its benefits. Metallic hydrogen might be more useful for gaseous-core engines than for chemical engines.

Much can be done in Earth-Moon space even with hydrogen-oxygen engines. Humans can return to the Moon and, as we have seen, a space shuttle–space station–lunar spaceship–lunar space station–lunar shuttle route would reduce the cost per kilogram of payload to far below what it was for the pioneering Apollo astronauts. The space station, space factory, space base, and lunar base are all within the reach of existing space technology.

The Space Shuttle itself may have an Earth-tied offshoot. For many years the idea of a rocket plane has intrigued aerospace engineers. Now we have such a rocket plane, the Space Shuttle. A modified Space Shuttle, flying at suborbital speed, could be a hypersonic long-distance airliner. Scheduled flights halfway around the Earth in little more than an hour would be revolutionary. For example, an American medical scientist could live in New York while doing research at the Pasteur Institute in Paris. Economical hypersonic flight would speed the erosion of the barriers of nationality that now divide humankind into over 150 compartments. An obvious problem would then crop up: reducing travel time to and from airports. Our scientist, for example, might take two hours to drive to Kennedy International Airport from his suburban home, fly to Orly Airport near Paris in 30 minutes, and take an hour to get to the Pasteur Institute.

There is no sense in waiting for the potential limit of the chemical rocket engine to be reached before starting work again on development of nuclear rocket engines. Years of progress and invaluable experience have been lost by the decision to end NERVA. The NERVA project (Nuclear Engine for Rocket Vehicle Application) should be reinstated, and the development of both solid-core and gaseous-core engines should be resumed as soon as possible. While gaseous-core engines have a higher potential specific impulse than solid-core engines, the solid-core engine will take less time to develop. A practical solid-core engine would have a specific impulse of about 800 seconds or about twice that of the hydrogen-oxygen engine. It could lower the cost of space flight in Earth-Moon space. The Space Shuttle could carry the stage (that is, the engine, its propellant tank filled with hydrogen, and the other equipment required) up to a parking orbit where it would be attached to the spacecraft brought up on a previous flight. The engine would be started in space and remain in space, thus overcoming objections about radioactivity.

Since 1973, gaseous-core engine studies have stimulated research on developing gaseous-core reactors for power plants and other uses on Earth. The gaseous-core or gas-fuel reactor (GFR) has several advantages over the solid-core reactors in use all over the world. The most common types of nuclear reactor are the pressurized-water reactor (PWR) and the boiling-water reactor (BWR). They have solid cores of uranium and are used for generating electricity. The main difference between them is that the PWR heats water under pressure to a higher temperature than the BWR. The water is heated to make steam, which drives a turbine, which drives a generator, which generates electricity. The core—the energy source—is made up, usually, of 40,000 fuel rods filled with a total of 100,000 kilograms (220,000 pounds) of uranium-238 enriched by 3.3 percent of uranium 235. The 3300 kilograms (7260 pounds) of potentially fissioning uranium-235 are the reactor's power source.

In contrast, the gas-fuel or gaseous-core reactor, still in the experimental stage, may need only 10 kilograms (22 pounds) or so of uranium-235 to operate,[6] only about 1/330th as much as a pressurized-water reactor of equivalent power, about 1000 megawatts or a million kilowatts. The uranium-235 is not in solid form. It is a hot gas that circulates continually through the reactor. As it circulates, part of the stream can be diverted, the radioactive byproducts removed, and the uranium-235 gas returned to the power cycle. In this way, the level of radioactivity in a gas-fuel reactor could be kept far below that of a pressurized-water reactor. Thus it is intrinsically much safer than solid-core reactors such as the PWR and the BWR. Since it does not have a solid core, it cannot have a core meltdown. If the gas-fuel reactor started to overheat, the pressure in the gaseous core would increase and relief valves could immediately bring the pressure down to a safer level, which might involve stopping the nuclear reaction.

Gas-fuel reactors could be designed to transmute the more dangerous radioisotopes such as plutonium, curium, and americum by neutron-induced reactions into less toxic isotopes of stable elements and so greatly diminish the problem of disposal of radioactive wastes.[7] In the long run, the best solution could be such "nuclear garbage disposals" that would get rid of long-lived radioactive wastes instead of storing them in abandoned mines where they would undergo lengthy natural decay.

Gas-fuel reactors could be designed to have thorium "blankets" lining their interiors. The thorium would be converted by the nuclear reactions into uranium-233. The uranium-233 would fission and simply replace the used-up uranium-235, keeping the reaction going. This type of GFR would be its own breeder. Also, it would avoid production of plutonium, unlike the conventional PWR, which runs on uranium-235. All in all the GFR could be the safest power plant ever built, safer than the PRW and, taking into consideration the overall dangers, safer even than conventional oil-fired, coal-fired, and gas-fired power plants.

Another advantage of the GFR is that it could be more efficient than the PWR, possibly converting 65 percent of its energy into electricity, thereby reducing both the cost of electricity and the volume of waste heat. Since the fissioning uranium in the gas-fuel reactor is a gas, not a solid that must be kept below the melting point, it does not have the temperature limitations of the pressurized-water reactor. It could be run at a very high temperature with a great increase in efficiency. A GFR would have constant refueling, whereas the PWR must be periodically shut down to replace the used fuel rods. And, of course, the cost of manufacturing tens of thousands of fuel rods for each reactor would be eliminated.

Phase 2 of physics takes in fusion research as presently envisaged—that is, up to the attainment of a commercial fusion reactor, which could be a magnetic confinement reactor or an inertial confinement reactor; possibly both types might be successful. There is a popular belief that fusion is almost impossible to attain and that, even if attainable, it will not be available for use for decades. This belief is unfounded. With strong support by Congress, fusion could become a commercial reality by the 1990s.

In view of its enormous potential benefits, fusion research has been kept on a starvation diet. For example, one of its most promising devices, the fusion torch (discussed in Chapter 6) has been neglected. The fusion torch could convert solid waste from a costly nuisance to an endless resource, make mining obsolete, and make recycling a routine function of the American economy.

Phase 3 of physics would center on mastering advanced fusion technology. All existing fusion research and development projects work with deuterium and tritium, since they are the isotopes that fuse most readily. Once a practical fusion reactor is realized, researchers will seek much higher temperatures, more powerful magnetic fields, higher plasma pressures, and more energetically favorable reactions. The outcome could be heavy-ion reactors operating at temperatures of a billion degrees Kelvin. They might convert almost all of their energy output directly into electricity, without creating any radioactivity.

The annihilation reactor is listed with a question mark. The complete conversion of matter into energy is possible only by the annihilation reaction. However, more energy is consumed in producing antimatter than is given off by the annihilation reaction. Whether or not this reaction can be made practical depends in large part upon whether economical techniques for producing antimatter can be developed. The one practical use of a solar-power satellite might be production of antihydrogen with solar energy.

Besides the central research fields, several related projects would be part of the proposed long-range space program. They would be supported by a Solar Neighborhood Institute devoted to assisting research on the region of space within 100 light-years of our solar system. The Solar Neighborhood Institute would be modeled on the Lunar and Planetary Institute of Houston, Texas, which conducts and supports research on the major bodies of our solar system. The new Institute would have four major programs, at least to start:

1. *Extrasolar Planetary System Search:*

The Institute would finance and coordinate projects on the design, construction, and operation of telescopes for detecting planets around nearby stars, as discussed in Chapter 14. It would also serve as a center for conferences of interested astronomers, optical scientists, astrophysicists, and engineers.

2. *Search for Extraterrestrial Intelligence:*

We cannot be certain that intelligent signals from space will never be received, for *never* implies eternity. Even if the search should take decades, it will be worthwhile, for the benefits to radioastronomy and electronics will be great.

3. *Physics of Nearby Stars:*

Within a few years we should be able to resolve the disks of nearby stars by improvements on existing telescopes—that is, by means of telescopes in space as well as new instruments. This will begin a new era in stellar astrophysics. At present we can see only the integrated properties of nonvariable stars, since they can perceived only as points. By use of theories, astrophysicists have been able to deduce the distribution of characteristics such as temperature and brightness over the disks of these stars, but this is not a substitute for direct observation. We will be able to better determine the relation of the Sun to other stars in the solar neighborhood. Because we see the Sun as a disk, rather than as a point, it is very difficult to accurately compare the Sun to other stars that have similar physical characteristics. Comparison of our solar system to other solar systems, which we hope to discover, will enable us to better understand the relation of planets to their parent star. This will help us place our solar system in context with other solar systems.

4. *Interstellar Flight Atlas:*
Our knowledge of our stellar neighborhood is very sketchy. We have only a rough idea of the number of stars within 100 light-years of the solar system. We know little about the precise location of these stars, their sizes, masses, composition, surface temperatures, magnetic fields, space motions, distances, and planetary systems, if any. The distribution of the interstellar medium in this volume of space is poorly known. The Interstellar Flight Atlas would be a continually updated compendium that would present a complete picture of the domain of space that starships could explore in the next century. Representing both pure science and applied astrophysics, it would grow out of the current efforts of astronomers to collect astronomical data files. This work is now centered at the Stellar Data Center in Strasbourg and at the NASA Goddard Space Flight Center.

Many astronomers, both professional and amateur, are fascinated by the cosmological problem—the question of the structure and evolution of the universe. But, of this, the average American has a more specific curiosity: Are there planets out there that we can explore and colonize? If we can answer this more limited question, then the resulting public support for space science will help us answer the broader one.

Notes

1. Harold B. Finger, General Manager, Center for Energy Systems, General Electric Company, private conversation.
2. Wilfred J. Mellors, Head, Washington Office, European Space Agency, private communication.
3. Francis C. Schwenk, Manager, Space Utilization Systems Office, Energy System Division, National Aeronautics and Space Administration, Washington, D.C., private communication.
4. Michael A. G. Michaud, "Spaceflight, Colonization and Independence: A Synthesis. Part One: Expanding the Human Biosphere," *Journal of the British Interplanetary Society*, 30 (1977), 83–95.
5. Robert L. Forward, "A National Space Program for Interstellar Exploration," Future Space Programs 1975, vol. VI, Subcommittee on

Space Science and Applications, Committee on Science and Technology, U. S. House of Representatives, Serial M, 94th Congress, September 1975.

6. H. Weinstein, T. S. Latham, M. Suo, E. Maceda, S. Chow, H. Helmick, "Reactor Concepts, Systems and Applications," in E. Krishman, ed., *Proceedings of the Princeton University Conference on Partially Ionized and Uranium Plasmas*, NASA, September 1976, p. 235.

7. J. D. Clement, J. H. Rust, A. Schneider, and F. Hohl, "Georgia Institute of Technology Research on the Gas Core Actinide Transmutation Reactor (GCATR)," in E. Krishman, ed., *Proceedings of the Princeton University Conference on Partially Ionized and Uranium Plasmas*, NASA, September 1976, p. 255.

19
THE SPACE ENTERPRISE

In Chapter 18 we discussed space planning and proposed a long-range space program. We look now at its implementation and some possible results. The key to stimulating the American space venture is expansion of the powers granted to NASA by Congress. The National Aeronautics and Space Act of 1958, the law authorizing NASA and assigning its functions, has been outmoded by the space agency's progress in space technology. It must be updated, giving the agency authority to take advantage of the vast new opportunities that have opened up.

More Authority for NASA

At least three amendments should be made to the Act. First, NASA should have a mandate to establish a permanent, continually staffed space station in low Earth orbit. This station is the critical transition point for full-scale exploitation of space. It is the "space Rubicon." The Soviet Union has had the Salyut-6, which is actually a space station, in orbit for more than a year. The United States has abandoned its space station, the Skylab, which has reentered the atmosphere and burned up, and has no plan to replace it.

Second, NASA should be assigned authority to plan on a long-term basis. It could adopt the program envelope technique of the European Space Agency or work out its own method. If the space agency could do such planning even for a decade at a time, teams of scientists and engineers could be better held together, improving skills and effectiveness with each new assignment, and the agency's steadier level of activity would yield increased efficiency.

Third, NASA should also be given a measure of financial independence. It should be allowed to charge, say, 5 percent more than the direct cost of the operation of the Space Shuttle to its users. The surplus would be set aside as a discretionary fund the Administrator of NASA could use for long-range planning, including analytical studies and advanced research and development. This would give the space agency some control over its future. As an example, Congress could authorize a program of over a decade to explore the solar system with unmanned spacecraft. NASA would be able to conduct preliminary planning and analysis with its own funds.

The agency should also be encouraged to maintain continued world leadership in space by international cooperation. Joint operations with the European Space Agency, Japan's National Space Development Agency, and other Free World countries interested in space, should be treated as a major concern, not as a minor issue of little importance.

A Free World Space Consortium

One way to nurture cooperation would be to establish a Free World Space Consortium. Although its title would be imposing, it could start very modestly as a small office in Washington, where representatives of the three agencies would meet frequently to work out plans for collaboration. At first, these representatives might have no authority to commit their agencies. They would report to their superiors, and agreements would have to be approved by the heads of the three agencies. The main distinction between the Free World Space Consortium and the present meetings on specific projects would be the recognition that international cooperation is essential to the future of the space enterprise and world democracy. As it progressed, other national and international space agencies would join, and it would become a federation of space agencies.

All agreements reached by the consortium or a similar organization would be for peaceful purposes only, and all its information should be completely open to the world's media. For its part, the United States could do this most effectively by completely separating its peaceful, open space missions from military and secret missions. The Department of Defense could be assigned its own Space Shuttle (or shuttles), which would be used only by itself and the intelligence agencies. The military space shuttle system could be based at Vandenburg Space Center, where it would have its own facilities, astronauts, and service personnel apart from those of the civilian shuttles using Vandenburg. Such an arrangement would facilitate international space cooperation by insuring its completely peaceful nature. In contrast, the Soviet space program is completely military. Also, on principle, if a government wants to keep some space missions secret, it should try to keep them completely secret and totally separate from the peaceful uses of space. It makes no sense to have degrees of secrecy.

The benefits to the American space program from international partnership would soon become apparent. NASA, with its partners, would be able to carry out projects that it has had to postpone for lack of funds. The space station is a case in point. An influx of space revenue would be the best answer to the argument that space is a waste of money. Once space becomes profitable, it may be expected that the enemies of space enterprises will do a fast somersault and begin to argue that the United States is "giving away" profitable and job-rich projects that should be kept in this country. The answer is, as President Kennedy said, "A rising tide lifts all the boats."

The Western Europeans have learned from long and hard experience the need for close cooperation in space. In December 1960, ten European nations agreed to form the European Space Research Organization (ESRO) to give European scientists their own access to space.[1] The European Launcher Development Organization (ELDO), which was to design, construct, and fire the rockets for launching satellites, was started in February 1964. The first European satellite was launched in 1968. A series of scientific satellites were sent into orbit. As the defects of separating the satellite program from the launcher program became evident, it was decided to merge the two to form a

European equivalent of NASA. The European Space Agency (ESA) came into existence on May 31, 1975.

Almost all of the Western European countries now belong to ESA: Belgium, Denmark, France, West Germany, Italy, the Netherlands, Spain, Sweden, Switzerland and the United Kingdom. Norway is taking part in one ESA program and Austria in two, and probably both will eventually become full members. Later on, the other Western European countries may join to take advantage of the economic and technical benefits of membership in a flourishing space federation.

ESA has designed and constructed the Spacelab, the manned laboratory that will be carried in the Space Shuttle for many of its missions. It is the most important international space project ever undertaken, and its success will be a strong stimulus for collective action in space. Forty Europeans firms shared in designing and building Spacelab. To them, the expertise they gained thereby is as important as the revenue.

ESA and its predecessor, ESRO, have over the past eleven years launched 14 satellites, all of which have operated successfully when they reached orbit. ESRO-II, for example, observed cosmic rays; ESRO-1B observed the ionosphere; HEOS-A1 the solar wind; TD-1 the optical and ultraviolet spectra of many stars; COS-B investigated hard X rays; and GEOS is observing the Earth's magnetosphere from geostationary orbit. Also, in the field of space science, ESA is developing EXOSAT, a satellite for the very precise location of X-ray sources, is making a significant contribution to NASA's Space Telescope, and is cooperating with NASA on the International Solar Polar Mission. In space applications, ESA has sent up Meteostat, which has been providing European and African meteorological services with information on the weather for several years, has had an experimental 11/14 GHz communications satellite working in orbit for more than a year, and is developing an operational European communications satellite system.

While three ESA satellites have been launched by NASA, the European Space Agency is developing its own rocket systems. In July 1973, the Europeans decided to build up their own launching capacity and so gain a share of the launching service market in the 1980s and the 1990s. Between 1980 and 1990, about 200 satellites are expected to be sent into geosynchronous orbit, so this will be a large industrial activity. ESA's entry will be the Ariane, the largest rocket undertaken so far by Western Europe.

Ariane is designed to launch 950 kilograms (2500 pounds) of payload into geosynchronous orbit, or 2500 kilograms (5500 pounds) into low Earth orbit. The rocket is 47 meters (115 feet) long and has a mass of about 207,000 kilograms (455,000 pounds). It has three stages. The first stage is propelled by four rocket engines that burn UMDH (unsymmetrical dimethyl hydrazine) with nitrogen tetraoxide to generate a thrust of 2,745,000 newtons (638,000 pounds). The third stage is a cryogenic system using liquid hydrogen as fuel and liquid oxygen as oxidant.

As the development of Ariane was decided in 1973 when the Space Shuttle program was well under way, the question arises why ESA went ahead with it instead of relying upon the Space Shuttle to launch its satellites. Wilfred J. Mellors,[2] the Head of the Washington office of the European Space Agency, explained that Ariane was undertaken for three reasons:

First, there was the question of political independence. The ESA nations were convinced that they must have guaranteed access to space and felt that, even though NASA had always been most punctilious in implementation of the U.S. launch service policy, circumstances could arise in which NASA, however much they wished to do so, would be unable to carry out a launch for a foreign customer. Consequently, Europe (like the USSR, Japan, and China) had decided that an independent launch-vehicle capacity was a political necessity. The second reason was political and technological stature, which today are by no means independent. It was considered essential that Europe demonstrate its competence in all fields of space activity, including space transportation. The final reason was the balance of payments. To date, ESRO and ESA have spent approximately $200 million on NASA launch services, all of which has come from the European taxpayer, and there was naturally a strong feeling that it would be better to spend this money in Europe.

Ariane will certainly be used for many years. The plan includes four development launches scheduled between November 1979 and October 1980, six operational vehicles in production with a further batch to be ordered, and firm plans for at least eight operational launches.

The designers of Ariane deliberately eschewed any technological "cleverness." They used only the accepted state of the art in rocket technology and manufacturing techniques. Development costs and risks were kept to a minimum, and production costs were held down. For example, the first and second stages employ the same fuel, oxidant and rocket engines. Ariane has done better than expected. It was designed to take 1500 kilograms (3300 pounds) into transfer orbit, but

ESA will now be able to guarantee a performance of 1700 kilograms (3740 pounds) into geostationary transfer orbit, which, because of the favorable latitude of the launch site, corresponds to a performance of over 1800 kilograms (4000 pounds) for a vehicle launched from Kennedy Space Center.

The European Space Agency is conducting studies on how Spacelab should be further developed and employed. Spacelab could easily become the nucleus of a space station. For its part, ESA is developing a European operational communications satellite system that will enable the people of the ESA countries to make telephone calls by satellite. ESA is also working on a maritime satellite program and is hoping to embark on an experimental television broadcasting system. The technological unification of Europe is keeping pace with its economic and political unification.

The European scientists have had a taste of what space science can do, and they want more space science missions. Among these projects being considered are a Grazing Incidence Solar Telescope (GRIST) to supplement NASA's proposed Solar Optical Telescope, an astronomical telescope operating in the far-ultraviolet region of the spectrum, and cooperation with NASA on a mission to Halley's comet and the Tempel II comet. Looking further, they would like to have ESA launch a Lunar Polar Orbiter satellite somewhat like the one that NASA has proposed but Congress has rejected. POLO (Polar Orbiting Lunar Observatory) would observe the Moon's magnetic field and determine whether it has a core like the Earth's. The Europeans also have made bolder proposals. One is to have Ariane launch a spacecraft that would speed through the asteroid belt between Mars and Jupiter, observing four asteroids as it flew past them. A second is to have Ariane dispatch a spacecraft to meet Halley's comet after it has swung around the Sun in 1986. A third is to cooperate with NASA on a Mars mission. The aim would be to place several inflatable balls on the surface of Mars. These balls would be blown up to 6 meters (20 feet) in diameter and roll over the terrain, making measurements as they rolled with the instruments they contained.

The Europeans have good reason to be treated as space partners. Western Europe's gross product is roughly equal to America's, while in many fields of science and technology, it is equally advanced, and in some fields, such as breeder reactors, Europe is ahead.

Are the Western Europeans willing to work more closely with the United States in space than they have been? Yves Demerliac, the Secretary-General of Eurospace, the organization of European companies interested in space, was asked for his views.[3] He replied:

I believe that European public opinion on space matters might well swing toward support for increased participation when Shuttle is operating successfully and routinely, simply through recognition of the widening capability gap between the USA and Europe that will become increasingly apparent.

With regard to prospective space developments, M. Demerliac stated:

I believe there will be a need for a vehicle that will provide access to the space environment in which space industrialization can be developed and exploited. This does not necessarily call for a manned vehicle such as Shuttle-Spacelab, as new industrial techniques, in the space environment, could be accomplished in automated laboratories, launched by advanced versions of Ariane or by heavy cargo carriers.

Nevertheless, for many new developments in space industrialization in the future, the presence of man will be essential, and I would support a European development in this field in conjunction with USA or any other space power, but only if such collaboration guaranteed free and unrestricted access, by Europe, to the environment and the benefits to be gained. In the absence of such guarantees, I would consider the development of both unmanned and manned vehicles should be conducted by Europe on its own.

European industry, evidently, is willing to cooperate with the United States in space on equitable terms. Japan, too, is interested in cooperating on space projects. The Japanese are keenly aware of America's present preeminence in space. They have been carefully following a farsighted program for gaining expertise and learning how to exploit space as a national resource. In May 1960, the Government of Japan established a National Space Activities Council, which reports directly to the Prime Minister.[4] The next few years were spent largely on laying the groundwork, setting up laboratories, and building a small space center. The National Space Development Agency of Japan (NASDA) was started in October 1969 and the work expanded.

The Japanese have designed, constructed, and launched several types of satellites that demonstrate increasing sophistication—engineering test satellites, ionosphere sounding satellites, experimental communication satellites, meteorological satellites, and experimental broadcasting satellites. Some of them were made with the help of

American technology. The first Japanese satellite, the Osumi, which had a mass of 24 kilograms (53 pounds), was launched on January 11, 1970. The Yuri, an experimental broadcasting satellite, which had a mass of 350 kilograms (770 pounds), was launched from Kennedy Space Center on April 8, 1978.

An even more telling indication of their advance in space technology has been the progress of Japanese launch vehicles. The first Japanese rocket with orbital capacity was the L-4S. It was 17 meters (52 feet) long and had a mass of 9500 kilograms (21,000 pounds) and three solid motor stages. The N-I, at present the largest Japanese launch vehicle, is 32.6 meters (107 feet) long, with a mass of 90,140 kilograms (198,000 pounds), and has three stages. In 1980 an N-II launch vehicle will be scheduled for its first test flight. The N-II is 35 meters (116 feet) long, has a mass of 135,000 kilograms (296,000 pounds), and has three stages. Development of the H-I, which will be Japan's main launch vehicle, is under way. The H-I will be larger than the N-II, have a hydrogen-oxygen main engine, and be able to place 550 to 800 kilograms (1210 to 1760 pounds) in geosynchronous orbit, or 3200 to 4800 kilograms (7000 to 11,000 pounds) in a 200-kilometer (125-mile) parking orbit. The H-I will be succeeded by the H-II, which will be still larger and more powerful.

Japan is constructing a firm base for its space enterprise. Since Japan is the third industrial power in the world and has a remarkable record for producing and selling abroad advanced-technology goods such as color television sets, cars, supertankers, pocket calculators, and much more, its commitment to space is not surprising.

Japan has a long-range space plan. Having completed the first stage, it is now engaged in the second, looking forward toward space activities at the end of the century. The third stage will start in the 1990s, when Japan expects to conduct space operations on a global scale. Over a period of 13 years, they plan to set up a communications satellite system, a broadcasting satellite system, and a navigation satellite system. Their scientists will be using astronomical observation satellites, meteorological satellites, and earth observation satellites. Looking farther afield, they plan to send spacecraft to explore the Jupiter-type (Jovian) planets and asteroids, and to send a probe to inspect Halley's comet. The Japanese are considering developing their own manned spacecraft, in addition to placing payloads on the Space Shuttle and cooperating in other ways with NASA, ESA, and other national and international space organizations.

Clearly, both ESA and NASDA are confident that space can

yield great returns. They have demonstrated their competence and are ready for an expanded effort in cooperation with NASA with one proviso—the United States must accept them as partners, not as subcontractors or customers of the Space Shuttle.

Tadahico Inada, the Washington Representative of NASDA, was asked his opinion on this question.[5] Japan is interested, he replied, in close cooperation with the United States. Two or three years of working together would not be enough for future space exploration. He believes that a long-range plan as well as long-range cooperation is necessary.

Cooperation between the three would have to be wholehearted and extend to every field of space technology, such as construction of a low Earth orbit space station, large space structures, chemical and nuclear engines, multipurpose satellites, and so on. Japan would even be interested in buying a Space Shuttle, Mr. Inada said, if the United States were interested in selling one.

Are ESA and NASDA really ready to become full-fledged partners of NASA? A Japanese study of the growth of their budgets is illuminating.[4] In 1971 Japan spent about 170 million yen a year on space, while ESRO, ESA's predecessor, was spending 220 million a year and NASA was spending, in yen, about 12,000 million a year. The Western Europe and Japanese efforts, including the individual space programs of West Germany, France, and Great Britain, amounted to about 12 percent of the American effort. By 1976 the picture had changed radically. Spending by the United States had declined to the equivalent of 10,322 million yen a year, while Western Europe and Japan, all told, were spending about 4600 million a year, or about 44 percent of the American expenditure. Since then, Japanese and European funding has been on the rise, while the American space budget has barely held its own. A joint space budget might easily increase to $8 billion a year in the first few years. Even more important, in the long run, would be the powerful effects of united backing of space activity by the most advanced countries on Earth with their huge scientific, technological, and industrial resources.

There would be another outcome from space growth: an increase in the number of spaceports for manned spacecraft. They might grow from the present single American spaceport and the two Russian ones to possibly a dozen by the end of the century. As space travel becomes routine, people will not be content to fly halfway around the globe, even in supersonic airliners, in order to catch a shuttle for the space station or the Moon.

Earthport Project

A proposal for space cooperation, widely different from the one just discussed, has been advanced by a private American nonprofit foundation, the Sabre Foundation.[6] It advocates establishment of Earthports, large spaceports at or near the equator. Earthports would be used for launching satellites and eventually manned spacecraft. They would also contain free trade zones. The aim of the project is to give many countries, especially the poorer ones, a share in space operations. Since many backward countries are located in the tropics, their location would seem to give them an advantage.

The favorable features of an equatorial location for a spaceport are apparent. Since the Earth rotates once in 24 hours, a satellite or spacecraft launched due east from the Equator gains an added 1600 kilometers (1000 miles) an hour over a launch from the regions of the North or South Poles. This gain in velocity varies, of course, with the latitude.

There are two additional advantages. When a satellite is launched from, say, Kennedy Space Center, and is to go into a geosynchronous orbit in the plane of the equator, it must make a "dog leg" turn to do so. This requires extra velocity and extra fuel. A satellite launched from a spaceport on the equator could go straight into a geosynchronous equatorial orbit. Also, a satellite launched due east from a site on the equator will pass overhead every time it comes around, so rendezvous with it requires a minimum of fuel.

Earthports, it is claimed, would be the first truly international space centers. Starting with satellite flights, an Earthport would progress to manned space flight, space industrialization, and solar-power satellites. Several tropical countries have shown interest in this scheme, and in time it may become a reality. Land in their free trade zones would be leased to industrial firms, bringing in to the first Earthport an estimated revenue of more than $300 million a year. Some of these earnings would be transferred to the host country as royalties.

This is an interesting proposal, but its benefits may be less than alleged. Placement of spaceports has been restricted by the deficiencies of the early space vehicles. These weaknesses are being overcome. The Space Age began with multistage, expendable rockets. They took off at high accelerations, shed stages as they climbed for space, and, in the case of the manned rockets, the capsule, the last stage, landed in the ocean. The whole scheme was wasteful, cumbersome, and expensive, but there was no other way to get started on space flight. Kennedy

Space Center, the American spaceport, was built on the southeastern coast of the United States so that rockets launched from it could drop their spent stages safely into the sparsely traveled South Atlantic.

That era is past. The Space Shuttle is the first of the new generation of space vehicles. While the Space Shuttle still drops off its external tanks and solid rocket Boosters, the latter float down on parachutes and are retrieved to be refurbished and used over again. The Space Shuttle's eventual successor, the SSTO (Single-Stage-to-Orbit), will be a one-stage spacecraft that will go into space and return to Earth as an entity. Nothing will be dropped off in flight.

The requirements for future spaceports will therefore come to resemble those for the great international airports. Future spaceports will not have to be situated on the western border of an ocean. The Russian space centers are in the midst of the vast Asiatic continent. Kapustin-Yar and Tyuratam are not far from the Black Sea. They are at about the latitude of Vienna. Plestesk is near the Arctic and is north of Stockholm. Neither their midcontinent locations nor their high latitudes seem to have hindered the Russian space program. Even though Siberia has a long coastline on the Pacific, the Russians have not constructed a spaceport there, probably because it would be thousands of kilometers from their main industrial centers. For the Soviet Union as for the United States, proximity to industrial and science centers is important in locating spaceports. Then, too, as the specific impulse and therefore the exhaust velocity of rocket engines increases, the "free" boost of the Earth's rotation will decrease in importance until it becomes insignificant.

Possession of a spaceport is not a sure route to space mastery. After World War II, when many of the backward countries became independent, often one of their first moves after gaining independence was to build an airport and establish a national airline. The effect of these innovations on their economies was usually slight. National airlines do little to promote industrialization. A country that lacks industry, a skilled work force, and a cadre of scientists and engineers cannot hope to climb into space, so to speak, by pulling on its own bootstraps.

World Cooperation in Space

One way to bring all the countries of the world, no matter whether advanced or backward, into space operations would be to follow the example of Western Europe and encourage the formation of continental space agencies. While an underdeveloped nation cannot enter space by itself, it could do so as partner in an international grouping. Nations could be helped to set up continental space agencies that would work with a NASA-ESA-NASDA federation. A Latin American Space Agency, an Asian Space Agency, and an African Space Agency would, with NASA, ESA, and NASDA, cover most of the world. In South and Central America several countries are advancing to the level of industrialization of Europe: Mexico, Brazil, Colombia, Venezuela, Chile, and Argentina. In Asia, India has begun its own space program, while the industrialized countries of the region, such as Australia, New Zealand, Taiwan, South Korea, Singapore, and Hongkong, along with India, might be interested in forming an Asian Space Agency. Africa, except for the Union of South Africa, is just beginning to industrialize. However, an African Space Agency would stimulate the growth of industry and science on that continent.

The future of space lies in international cooperation. Once manned space missions create a large number of openings for astronauts, astronaut candidates should be recruited, not only from the United States, Western Europe, and Japan, but from other interested countries as well. Having "their" astronauts going on space missions would be a source of local pride. There should also be arrangements for bringing scientists, engineers, and technicians to institutions in the United States, Europe, and Japan for advanced training in their fields of space science and space technology. The continental space agencies could be assigned projects that would be subcontracted to companies, universities, and technical institutes in their member countries. Initially, these assignments would be made as much for the on-the-project training they afforded as for their contribution to the space enterprise. They would be closely monitored until the subcontractors had demonstrated their competence.

Just as there are no barriers in space, there should be no barriers between human societies. In the long run, the greatest contribution of space could be the fostering of world peace.

Amateur Space Societies

There is today an obvious gap in space activities, at least in the United States. Almost every cause in our country has its organized supporters. One can hardly find a single pursuit, no matter how rare or esoteric, that does not have its organization of devoted followers. It is astonishing that space, which arouses universal interest and encompasses almost every scientific and technical field, as well as many of the arts and humanities, is not well represented in our amateur societies.

The news on space events is usually given very scanty coverage by the media and barely hints at the fascinating features of the discoveries and developments made. The mission of Voyager I is a good example. The revelation of what is practically a miniature solar system of bodies circling Jupiter was reported for a few days and then dropped. The mission of Voyager II was almost ignored. Amateur space societies should be formed to keep up with the space adventure and to understand and appreciate it more fully. These societies could sponsor talks by scientists and engineers, hold discussion meetings and workshops, show films on space, and carry on other interesting activities.

These societies could be modeled on the excellent British Interplanetary Society, whose membership is made up of scientists, engineers, and people from other walks of life who are united by their keen interest in space. This wide base ensures the society the technical expertise it needs to be effective. The space societies should offer a reputable journal as a benefit of membership, such as the *Journal of the British Interplanetary Society* and the Society's popular space magazine, *Spaceflight*. Also, these societies should consider cooperating with the British Interplanetary Society, which has an international membership.

We have reviewed the status and potential of the space enterprise. The exploration and colonization of the solar system will be highly rewarding, and the possibility of interstellar exploration looks promising. However, the future of the space enterprise depends as much upon popular support as upon the achievements of the scientists, engineers, and technicians. By offering such support, the reader can participate in realizing the wonderful possibilities of space.

Notes

1. *Space—Part of Europe's Environment*, European Space Agency, May 1977, pp. 3–10.
2. Wilfred J. Mellors, Head, Washington Office, European Space Agency, private communication.

3. Yves Demerliac, Secretary-General, Eurospace, private communication.

4. *NASDA 78–79*, National Space Development Agency of Japan, 1978.

5. Tadahico Inada, Washington Representative, National Space Development Agency of Japan, private communication.

6. *Earthport Project, Status and Goals* (Santa Barbara, Calif.: Spring 1979), Sabre Foundation, pp. 9–12.

List of Acronyms

ARPA—Advanced Research Projects Agency
BWR—Boiling-Water Reactor
CELSS—Controlled Ecology Life Support System
ELDO—European Launcher Development Organization
EMPIRE—Early Manned Planetary-Interplanetary Round Trip Expeditions
ERM—Earth Return Module
ESA—European Space Agency
ESRO—European Space Research Organization
ET—External Tank
GFR—Gaseous-Fuel Reactor
GSO—Geosynchronous Orbit
HLLV—Heavy-Lift Launch Vehicle
LEO—Low Earth Orbit
MAP—Multichannel Astrometric Photometer
MEM—Mars Excursion Module
MMM—Mars Mission Module
NASA—National Aeronautics and Space Administration
NASDA—National Space Development Agency of Japan
NERVA—Nuclear Engine for Rocket Vehicle Application
OPEC—Organization of Petroleum Exporting Countries
OTV—Orbital Transfer Vehicle
PWR—Pressurized-Water Reactor
RAIR—Ram-Augmented Interstellar Ramjet
RIFT—Reactor-in-Flight-Test
SETI—Search for Extraterrestrial Intelligence
SPS—Solar-Power Satellite
SSME—Space Shuttle Main Engine
SSTO—Single-Stage-to-Orbit (spaceship)

INDEX

aberration, of starlight, 225–26
Abt, Helmut, 234, 238
accelerator-mass spectrometer experiments, 152–53
accelerometers, 226
Acidalia Planitia (Mars), 144
Advanced Research Project Agency, 189–90
AEC (Atomic Energy Commission), 66, 69, 70, 73
Aerojet General Corp., 67
Aeronautics Corp., 133, 134
Africa, space programs in, 320
aging, in interstellar travel, 188
Agnew, Spiro T., 134, 135
air pollution, 84, 86
Aldrin, Buzz, 3
algae, 14
   for atmosphere modification, 162
   on Mars, 149
Allen, Charles, 142
Alpha Centauri star system, 20, 192, 244, 253, 262, 270
   Alpha Centauri A, 217, 220, 286–89
   Alpha Centauri B, 217, 220, 286
   manned expedition to, 183, 184, 186–87, 201, 228, 263, 272–89, 299–300
   Proxima Centauri, 55, 217, 286
   triple-star configuration, 217, 286–89
      distances between stars, 221, 286

amateur space societies, 320–21
Ames Research Center (NASA), 20, 49, 148
amino acids, 14, 17
Anders, Edward, 17
Anderson, Claude, 13
Anderson, Sen. Clinton, 69
Andromeda galaxies, 175, 179
angular resolution, in stellar parallax, 218–19
Animal Colony, on starships, 256, 277–78
animal tissue, for planetary colonization, 258
annihilation-powered starships, 203, 205–7, 302t
annihilation reactors, 203, 205–6, 302t, 306
antennae, for interstellar communication, 269–70
antimatter, 203, 205, 306
Antoniadi (astronomer), 129
aphelion, 117
apodized mirror telescope, 242, 245
Apollo class asteroids, 163, 164, 165
Apollo Mission (1969), 3, 42, 43, 44, 69, 75, 76, 134, 150, 268, 290, 291, 300–1, 303
   payload, 60–61
   rocket engines, 57, 69
   travel time, 122
Apollo-Soyuz mission, 42, 43, 292
archaeology, of extrasolar planets, 279–80, 281–82
Arecibo radiotelescope (Puerto Rico), 13
argon, in Martian atmosphere, 131–32

Argyle, Edward, 14, 15, 16, 236
Ariane rocket (ESA), 312–14, 315
Arizona Meteor Crater, 39
Armstrong, Neil, 3
Arrhenius, Svante, 10
Asia, space programs in, 320. *See also* Japan; China
asteroid belt, 162–63, 314
asteroids, 27, 31, 52, 162–65
   chemical composition, 164
   classes, 164
   exploration of, 164–65, 295
   moons of, 163
   orbits, 163
   origin, 36
astronauts, 263–67, 268, 292, 320. *See also* starships: crews
astronomical unit (AU), 29, 216
astronomy, 5
Atlas-Centaur rockets, 58, 61–62
atmospheres, of planets, 27
   modification of, 148–50, 160
atom bomb propellants, 188–89
atomic clocks, 237
Atomic Industrial Forum, 85

Barnard's star, 196, 197, 230–33, 238, 250, 299, 300
   planets around, 231–33
base line, in stellar parallax, 215, 217, 219
beamed-power spacecraft, 191
Becquerel, Paul, 11
Bekey, Ivan, 92
Beichel, Rudi, 63
binary stars, 178–79, 229, 237
   as gravity machines, 193
Black, David C., 232
black dwarf stars, 230
black holes, 4, 78, 277
blackbody radiation, 177
blue shift, in starlight, 224. *See also* Doppler effects
BMEWS (Ballistic Missile Early Warning System), 19
Boeing Company, 64, 85
boiling-water reactor (BWR), 304
Bond, Alan, 196, 300
booster engines, for nuclear rockets, 135–36, 136 *illus.*, 137 *illus.*
boredom, on interstellar flights, 268
boron-sail brake, on starships, 200, 201
Bracewell, Ronald, 243
British Interplanetary Society, 196, 262–263, 321
Burroughs, Edgar Rice, 130
Bussard, Robert, 182, 198–99

calcium loss, during space travel, 43, 44, 45, 47
Callisto, 33, 157, 166, 167–68, 168 *illus.*, 169
   as base for exploration, 168–69
   as starship building site, 263
carbon, carbon compounds, 7, 34, 133
carbon dioxide:
   in Earth's atmosphere, 83

(carbon dioxide, *cont.*)
   on Mars, 131, 132, 142, 151
   on Venus, 162
carbon-14, 132
carbon monoxide, in motor vehicle exhaust, 84
carbonaceous chondrites, 17, 164, 165
Carr, Michael A., 142
catalytic nuclear ramjets, 201–2, 204 *illus.*, 251–53
   mass requirements, 251–52
celestial coordinates, 110–12
celestial equator, 111 *illus.*, 112
celestial sphere, 110
   mapping of, 220, 268
   star positions, changes in, 213
CELSS (Controlled Ecology Life Support System), 49–53, 91, 102, 255–56, 266, 269, 282
centrifuge experiments, 47, 48
Ceres, 162, 165
Chappell, W. R., 12
Charon, 30, 172
chemical propellants, engines, 55–56, 58, 63, 295, 301, 303
children, on interstellar flights, 264, 265, 268, 276, 277
China, People's Republic of, space programs, 82, 313
Chiron, 163
chlorine gas, U.S. production of, 86
Christy, James W., 172
Clark, T. A., 243
Clarke, Arthur, 5, 184
clocks, in interstellar navigation, 226, 228
CNO bi-cycle, 202–3
coaxial-flow reactors, 71–72
Coblentz (astronomer), 129
Cocconi, Giuseppe, 18
Cohen, Bernard L., 85
comets, 27, 177, 194, 314
   origin, 36
   periods, 165
   physical properties, 33, 165
      organic compounds, 16–17
communication:
   interstellar distances, 253–54
   Solar System, 253
   See also space communication systems
communications satellites, 82, 92, 312, 314, 316
   life spans, 93–94
composite materials, for starship construction, 252
constellation, 213, 214 *illus.*, 215 *illus.*
coordinates, coordinate systems, 105–15
   ecliptic, 112–113, 114 *illus.*
   equatorial, 110–12
   galactic, 211 and *illus.*, 212–13
   planetary, 105–10
COS–B satellite, 312
cosmic rays, 38, 41, 42, 156, 157
   effects on retina, 48
Cosmos biological experiments (1975, 1977), 47
Crick, Dr. Francis, 10–12, 24*n*

Danielson, R., 241
Darwinian evolution, 15–16
declination, 111 and illus., 112
Deep Space Network (NASA), 269
Deimos, 33, 110, 128
Delta Pavonis, 262
Delta rockets, 58, 61
Demerliac, Yves, 314–15
Devereux, W. P., 227
directed panspermia theory, 11–13
distance determination. See interstellar navigation
Dole, Stephen, 177, 234
Doppler effects, 19, 192, 269
 in starlight, 224
 and proper motion, 224, 225 illus.
 and velocity in interstellar flight, 227–28
double stars. See binary stars
Douglas Aircraft Corp., 133
deuterium, 13, 199, 201–2. See also nuclear fusion; rocket engines
Drake equation, 9–10, 17
Drake, Frank D., 9, 18–19, 22
Dual Expander Engine, 63
dust grains, in star formation, 234–35
Dyson, Freeman J., 189, 190, 193

Earth:
 atmosphere, 29, 254
  air pressures, 37–38
 climate, 78, 81, 83
 coordinates, 108–9, 109 illus.
 gravitational field, 29, 56
 life on, origin of, 14–17
 life-support system on, 254
 orbital plane, 113
 physical properties, 28t, 29
 surface area, 3
Earth-Moon space, 4, 75
 colonization of, 75–79, 303. See also space colonies
Earth Orbit Space Station, 75, 98–99
Earthport Project, 318–19
eclipses, of extrasolar planets, 236
ecliptic coordinates, 112–13, 114 illus.
educational facilities, on starships, 264, 268
Ehricke, Krafft A., 4, 169, 194, 300
Eichhorn, Heinrich, 232
Einstein, Albert, 203
Eldred, Charles, 61
elliptical orbits, 116–17
EMPIRE (Early Manned Planetary-Interplanetary Round-trip Expeditions), 133–34
energy consumption, and living standards, 81–82
Epic of Gilgamesh, 260
Epsilon, Eridani, 18, 20
equators:
 celestial, 111 illus., 112
 galactic, 211 and illus., 212
 terrestrial, 108, 109 illus.
equinoxes, 111 illus., 112, 113

ERM (Earth Return Module), 134
Escherichia coli bacterium, 15
ESRO (European Space Research Organization), 311, 312, 313, 317
 satellites, 312
Europa, 33, 166 illus., 167
European Launcher Development Organization (ELDO), 311
European Space Agency (ESA), 82, 89, 292, 310, 316
 founding (1975), 312
 program envelope technique, 293–94, 310
 rocket systems, 312–13
 satellite program, 312
 Spacelab studies, 314
Eurospace, 315
evolution, and gene mutation, 15–17
exercise machines, for interstellar travel, 259
exhaust velocity, of rocket engines, 56–58, 65, 124, 319
exobiology, 13–14
EXOSAT, 312
expansion ratio, of rocket nozzles, 60
Explorer I, 39–40
External Tanks, 87, 88–90
extrasolar planets, 229–45, 261–62, 298, 299–300, 302t, 306. See also Alpha Centauri; Barnard's star
 atmospheres, 282
 detection methods, 235–45
 habitable planets, frequency of, 280
 manned exploration of, 273, 279–83
 mapping of, 280–81
 mass ranges, 230
 in multiple star systems, 233–34
extraterrestrial intelligence, search for, 17–24, 78, 279–85, 306
extraterrestrial life, search for, 6–17, 78, 279–85, 287–89

Fabry-Perot radial velocity spectrometer, 239 illus.
Fennelly, Alphonsus, 200, 236–37, 241–42, 243, 300
Fig Tree algae, 14
Finger, Harold B., 70, 292
flare stars, 286
Fletcher, James C., 77
flyby spacecraft, 294
flying saucers, 283–84
food supplies, in space travel, 49, 50, 52, 254, 255–56
Forward, Robert L., 192, 298–300
Fourier-transform spectroscopy, 238
Free World Space Consortium, 310–11
Frye, G., 241–42
fusion reaction. See nuclear fusion
fusion torch, 164, 305

galactic equator, 211 and illus., 212
galaxies, 4, 179, 244. See also Milky Way Galaxy.
 as reference points for navigation, 213
Galileo, 6
Galileo probe, to Jupiter, 8

Ganymede, 13, 33, 157, 167
gas chromatograph-mass spectrometer experiments, 133
gaseous-core engines, reactors, 70–73, 124, 139–40, 145, 302t, 303–4
gas-exchange experiment (Viking), 132
gas-fuel reactor (GFR), 302t, 304–5
Gatewood, George, 232, 233, 239–40, 243, 244, 245
Gemini mission, 3, 42, 43, 60
gene mutation, and evolution, 15–17
General Atomics Co., 189
General Dynamics Corp., 133
"General Limits of Space Travel, The" (von Hoerner), 181–82
geochronometry, 281–82
GEOS (satellite), 312
geosynchronous orbit, 61, 62, 318
Glaser, Peter, 84
glass production, at lunar bases, 96
Goddard Space Flight Center, 307
Goldreich, Peter, 234
graphite composites, for starship construction, 252
Grazing Incidence Solar Telescope (GRIST), 314
gravitational forces:
    Earth, 29, 56
    in Solar System navigation, 116
gravity, 116
gravity machines, for interstellar flight, 193–94
great circle, 108
Great Red Spot (Jupiter), 166 *illus.*
Great Spiral Galaxy (M-31, Andromeda), 174–75, 179, 220
Greek astronomy, 110
Greenwich meridian, 108, 109 and *illus.*
G-type stars, 238, 277
Gualtieri, Devlin M., 12–13
guidance system, in interstellar navigation, 210
gyroscopes, 226

Hall, Asaph, 128
Hall Crater (Phobos), 146, 147 *illus.*
Halley's comet, 314
heavy-element density, in star formation, 11
heavy-ion reactors, 302t, 306
Heavy Lift Launch Vehicle, 64, 80, 84
helioids, 169
Henry, Beverly, 61
Henry the Navigator, 2
HEOS-A1 satellite, 312
Heppenheimer, T. A., 196, 234
Herculina 532, 163
Hessberg, Rufus R., 45, 48, 52–53
hibernation, in interstellar flight, 187–88
high-energy orbits, 119 and *illus.*
Hoag, David G., 210, 226
H-I, II launch vehicles (Japan), 316
Horswell, C. J., 283
hospitals, on starships, 264–65
hour circles, 111 *illus.*, 112
Hoyle, Fred, 5

Huang, Su-shu, 238
Hunter, Maxwell, 185, 300
Huygens, Christian, 128
Hyde, R., 195
Hyde-Wood-Nuckolls engine, 195–96
hydrogen:
    as automobile fuel, 86
    in interstellar space, 182–83, 202
    as rocket fuel, 58, 65–66, 199–202, 251
        antihydrogen, 206
        metallic hydrogen, 58, 303
hydrogen bomb propellants, 189, 190
hydrogen extraction, at lunar bases, 88, 96–97
hydrogen-fluorine rocket fuel, 58
hydrogen-oxygen rocket engines, 58, 63, 303
hypothermia, 187

Iapetus, 171
I.A.U. 308 crater (Moon), 95 *illus.*
imaging interferometry, 240–41, 245
Inada, Tadahico, 317
India, space programs, 320
inertial confinement fusion reactors, 184–85, 195
infrared instruments, for detecting extrasolar planets, 243
inertial instruments, in interstellar navigation, 226, 228
interferometry, 236–37, 240–43
    long baseline, 236–37, 243
Interim Upper Stage rockets, 62–63, 80
International Solar Polar Mission, 312
interplanetary voyages, 70, 121–27. *See also* Solar System navigation
interstellar communication, 253–54, 269–70
interstellar exploration programs, 295, 298–300, 302t
Interstellar Flight Atlas, 226–27, 307
interstellar navigation, 210–18
    distance determination, 213–20
        angular resolution, 218–19
        base line, 215, 217, 219
        parsecs, 216–17
        stellar parallax, 214–15, 216, 217 *illus.*, 218–19
    galactic coordinates, 210–13
    inertial instruments, 226, 228
    motion measurements, 221–28
        aberration of starlight, 225–26
        radial velocity, 221–22, 223, 224
        relativity effects, 213, 226–28
        tangential velocity, 223–23 and *illus.*, 224
    reference points, 213
    time dilation, 228
interstellar probes, 295, 298
interstellar ramjet, 186, 197, 198, 250–51, 301, 302t
interstellar space, 3–4, 176–77
    density of matter, 176
    nebulae in, 176–77
    organic compounds in, 16–17
    temperature, 177
interstellar travel, 4–5, 180–91, 298. *See also* starships
    destination, choice of, 261–62

(interstellar travel, *cont.*)
    feasibility, 180–87, 261
    hibernation in, 187–88
    length of missions, 264–65
    time dilation in, 264
Io, 33, 166 and *illus.*, 167
    physical properties, 166–67
    volcanoes, 166
ion scoops, for hydrogen extraction, 96, 200–2, 252–53
Isaacman, Richard, 234, 235

Japan:
    nuclear power plants, 85
    space programs, 82, 310, 313, 316
        budget, 317
        launch vehicles, 316
        satellites, 315–16
Jensen, Oliver G., 232
Jet Propulsion Laboratory, 206
Jovian planets. *See* Jupiter; Saturn; Uranus; Neptune
J-2 rocket engine, 59–60, 60*t*
Jukes, Thomas H., 12
Jupiter, 6, 19, 27, 36, 99, 116, 123, 166 *illus.*, 220, 295
    gravitational force, 116
    life on, 8
    manned voyages to, 123, 124*t*, 126
    moons, 32–33, 52, 123, 166–67. *See also* Io; Ganymede; Callisto; Europa
        as exploration bases, 157, 166–69
        modification of, 169
    physical properties, 28*t*, 31–32
    radiation belt, 32, 167, 230
    ring, 32

Kapteyn's star, 222, 223
Kapustin-Yar space center (Soviet Union), 319
Kenknight, C. E., 242, 245
Kennedy, Pres. John F., 291, 311
Kennedy Space Center, 318–19
Kepler, Johannes, 35, 116–17, 176, 231
Khrushchev, Nikita, 40, 291
Kitt Peak National Observatory, 245
Kiwi nuclear rocket reactors, 67–68
Kowal, Charles, 163
Kramer, James J., 70
krypton, 132
K-type stars, 21, 177
Kumar, Shiv, 229

labeled-release experiment (Viking), 132
Lalande 21185, 300
Lampland (astronomer), 129
Langton, N. H., 39
laser fusion reactors, 184, 190–91
laser fusion rocket engines, 184–85
laser-powered starships, 191–93, 250
laser ranging, 115

Latin America, space programs, 320
latitude, longitude:
    celestial, 113
    galactic, 211 and *illus.*, 212
    terrestrial, 107 and *illus.*, 108, 109 and *illus.*
launch vehicles, 313, 316
law of universal gravitation (Newton), 116
laws of motion in orbit (Kepler), 116–17
laws of nature, 4–5
Lawton, A. T., 243
least-energy orbits, 117, 118 *illus.*, 119
Levy, Saul G., 234, 238
Lewis Research Center (NASA), 66, 71, 191
life:
    chemistry of, 7–9, 12–13, 16–17
    origins of, 13–17
    outside Solar System, 9–13. *See also* SETI
    search for. *See* extraterrestrial life
    within Solar System, 7–9
life support systems, 49–53, 292
    food supplies, 49, 50, 52, 254, 255–56
    ground-based systems, 51–52
    interstellar travel, 188, 254–56
    in space stations, 87–88
    waste management, 50–51
light, speed of, 184, 213
light-year, 184, 216–17
lignification of plants, and space travel, 49
LINAC (Linear Accelerator Craft), 200–1
livestock, on starships, 256
Local Group (galaxies), 4, 179
Lockheed Missile and Space Co., 67, 133
long-focus photometric astronomy, 230, 232–40, 243
longitude. *See* latitude, longitude
Los Alamos Linear Accelerator, 205
Lowell, Percival, 8, 128
Lowell Observatory (Ariz.), 128–29
L-type chondrites, 165
Lunar and Planetary Institute (Houston), 306
lunar bases, 94–102, 263
    guided tour of, 97–102
    industrial ventures, 95–97
Lunar Orbit Space Station, 75, 88, 263
Lunar Polar Orbiter satellite, 314
Lunar Spaceship, 75, 98–100

McAlister, Harold, 237, 244
McBride, J. P., 85
McElroy, Robert D., 148
McLafferty, George, 72
MacPhie, Robert H., 243
Magellanic Clouds, 4
magnetic fields, 156, 157
magnetic-intake fusion reactors, 199–200
main-sequence stars, 177
Mandeville, Sir John de, 2
Manned Orbital Transfer Vehicle, 63
Mare Imbrium (Moon), 127
Mariner spacecraft, 8, 55, 129–31, 161 *illus.*, 194–195

Mars, 27, 36, 128–53, 220, 285, 314
  atmosphere, 31, 130, 131 and *t*, 132
    modification plans, 148–50
  canal theory, 8, 128–29
  life on, 8, 130, 132–33, 150–51
  manned voyages to, 55, 70, 126, 128, 133–53
    costs, 139, 141
    landing, 137, 138 *illus.*
    NASA plans, 133–35
    rocket engines, 135–36, 135 *illus.*, 137 *illus.*, 139–40
    sketch of, 145–48
    travel time, 123, 124*t*, 125
    von Braun plan, 135–41
    weight history, 137 and *illus.*, 140–41
  navigation to, 115–17
    high-energy orbits, 119 *illus.*
    least-energy orbits, 117, 118 *illus.*
  observation history, 8, 128–29
  physical properties, 28*t*, 31, 129–31, 141–42
    canyons, 142, 144 *illus.*
    craters, 129–31
    mountains, 141–42, 152 *illus.*, 153 *illus.*
    polar caps, 131, 142, 143 *illus.*, 149, 151
Marshall Space Flight Center (NASA), 66, 67, 89, 133, 134
Martin, Anthony, 197, 199, 243, 300
Martin-Marietta Corp., 64
Marx, G., 192
mass-driver rocket engine, 57
mass ratio, of rocket engines, 57–58
Matloff, Gregory L., 186, 200, 236-37, 241–42, 243, 300
matter-antimatter annihilation engines, 181, 203, 205–7
Meglen, R. R., 12
Mellors, Wilfred J., 313
MEM (Mars Expedition Module), 134, 138, 139
Menzel, Donald, 283
Mercator projection, 110
Mercury, 8, 13, 23, 27, 117, 127, 157, 158 and *illus.*, 159–60, 295
  meridians, 109
  physical properties, 28*t*, 30, 157–58
  rotational periods, 158–59
Mercury mission, 3, 43, 60
meridians, 108–9 and *illus.*
Messier 31 galaxy. *See* Great Spiral Galaxy
metal extraction, at lunar bases, 101
meteoroid satellites, 40
meteoroid shields, 40
meteoroids, 27, 31, 125
  and space travel, 38, 39–40
meteors, meteorites, 39–40, 163
  organic compounds in, 17
Meteostat satellite, 90 *illus.*, 312
Michaud, Michael A. G., 294–98, 300
micrometeoroids, 31, 40
microfiche, in starship libraries, 257, 258

microwaves:
  in annihilation-powered engines, 206
  and detection of extraterrestrial life, 20–21
  electric power transmission, 80
  radiation, 82–83
Milky Way Galaxy, 3, 4, 174–79
  age, 11
  civilizations in, 9–10, 17–18
  coordinate system, 211–13
  Galactic Center, 174, 175
  halo region, 175
  life-suitable planets in, 10
  spiral arm pattern, 175–76
  star formation in, 9–11
minor satellites, 163
MMM (Mars Mission Module), 134, 138, 139
Model One space colony, 79–80
molybdenum, distribution of, in universe, 11–12
Moon (Earth), 3, 6, 33
  as communication base, 269
  as a fuel source, 88, 95–97
  limb of, as occulting edge, 242
  magnetic field, 314
  manned voyages to, 75–76, 122–23, 124–25. *See also* Apollo mission; lunar bases
  physical properties, 30
  as starship building site, 263
  surface features, 30
moons, in Solar System, 27, 169–71, 295
  of asteroids, 163
  as exploration bases, 157, 169–71
  orbits, 115–16
  origin, 35
  physical properties, 32–33
Morgan, David L., 206–7
Morrison, Philip, 18
Moskowitz, S., 227
motion sickness, during space travel, 44, 45
M-31. *See* Great Spiral Galaxy
multichannel astrometric photometer (MAP), 240, 244–45
  with space reflector, 243–44
multiple-star systems, 178–79, 229, 230–34. *See also* Alpha Centauri
  orbiting patterns, 233–34
  unseen companions, detection of, 230–31
Murchison meteorite, 17
muscle shrinkage, during space travel, 44

NASA (National Aeronautics and Space Administration), 156, 252, 316
  Astronaut Corps, 263
  Earth-Moon space colonization, 76–79
  launching services, 312–13
  life-support research, 49–50
  Mars explorations, 133–35
  nuclear rocket program, 66–67, 73
  program discontinuity, 70, 73, 290, 292, 293
  and Project Orion, 189–90

(NASA, *cont.*)
  space station studies, 87
  space planning, 291–92, 309–10
National Academy of Sciences, 291
National Aeronautics and Space Act of 1958, 309–10
National Radio Astronomy Observatory (Green Bank, W. Va.), 18
National Research Council, 91
National Science Foundation, 266
National Space Activities Council (Japan), 315
National Space Development Agency (NASDA—Japan), 310, 315, 316–17, 320
navigation, 210. *See also* Solar System navigation; interstellar navigation
Near East archaeology, 259–60
Near Stellar Network, 269, 276
near-stellar space, 4
nebulae, 176–77
neon-sodium cycle, in catalytic nuclear reactors, 203
Neptune, 8, 13, 27, 36, 167, 295
  moons, 157
  physical properties, $28t$, 32
NERVA (Nuclear Engine for Rocket Vehicle Application), 67, 68 and *illus.*, 69, 71, 140, 197, 292, 303
neutrinos, 156, 202
neutrino detectors, 156
Newton, Isaac, 43, 56, 116
nitric acid-hydrazine rocket fuel, 58
Nixon, Pres. Richard M., 69–70, 134, 135
N-I, II launch vehicles (Japan), 316
North Pole. *See* poles
NOVA rocket engine, 133–34
NRX-A2 nuclear rocket reactor, 68
Nuckolls, J., 195
nuclear fission process, 65, 70–71, 72, 185
  on extrasolar planets, 281–82
  in gas-fuel reactors, 305
  for propulsion, 194–95
nuclear fusion, 27, 34, 184, 190, 299, 301
nuclear fusion engines, 124, 156, 181, 184–85, 197, $302t$
nuclear fusion reactors, 87, 101, 156, 184, 199–200, $302t$, 305
  magnetic-intake, 199–200, 305
nuclear fusion starships, 194–96, 250
nuclear-light-bulb reactors, 71, 72
nuclear power plants, 81, 84, 85
  cost-effectiveness, 85
  high-temperature, 86–87
  at lunar bases, 96
  safety of, 85–86
nuclear pulse rockets, 186, 197–98
nuclear rocket programs, 65–73, 295, 298, 301, $302t$, 303–5
  booster engines, 135 and *illus.*, 137 *illus.*
  experimental reactors, 67–69
  fuel, 65–66, 70
  gaseous-core reactors, 70–73

(nuclear rocket programs, *cont.*)
  program phases, 72–73
  refractories, 66, 70
  regenerative cooling, 66
  solid-core engines, 66–67, 70
Nuclear Rocket Development Station (Las Vegas, Nev.), 66, 67, 70, 73
nutation, of plants, 48–49

Oak Ridge National Laboratory, 85
Oberon, 171–72
occulting disks, for extrasolar planet detection, 241–42
ocean sailing, 1–2
oil, world dependence on, 82
Oliver, Bernard M., 20–21, 183
Olympus Mons (Mars), 127, 141–42, 146, 151–52, 152 *illus.*, 153 *illus.*
O'Neill, Gerard K., 57, 79, 80–81, 164, 186, 200, 242, 300
Öort comet cloud, 165, 261, $297t$, $302t$
OPEC, 82
Öpik, Ernst J., 130, 182–83
Orbital Transfer Vehicle, 59, 94
orbits:
  of planets, 115–16, 117
  of spacecraft, 119–20
organic compounds, in interstellar space, 16–17
Orgel, Leo M., 10–12
Orion Nebula, 182
Osumi satellite (Japan), 316
oxygen extraction, at lunar bases, 95–96
oxygen-hydrazine rocket fuel, 258
"Outlook for Space, The" (NASA), 77–78

paleontology, of extrasolar planets, 279, 280, 281
Pallas, 165
Palmer, E. W., 283
panspermia theory, 10–13
parallax, 114 and *illus.*, 214. *See also* stellar parallax
parallels, 110
parsec, 216–17
payloads, of rockets, 60–61
  cost trends, 60–61, $61t$
perihelion, 117
period of rotation, 108
permafrost, on Mars, 142, 144
Phobos, 33, 52, 110, 128, 146, 147 and *illus.*
  as space station, 141, 146, 153
Phoebe, 171
photoelectric photometers, 163
photons, 193
photon thrust engines, 181–82, 192–93, 197
Pioneer spacecraft, 55, 57, 176, 184
plagioclase, 96
planetary coordinates, 108–10
planetary engineering, 148–53, 162, 169. *See also* atmosphere modification
planetesimals, 234–35
planet formation, 35–36, 38, 234–35

planets, of Solar System, 27–33. *See also* individual planets; moons; extrasolar planets
  atmospheres, 27
  manned exploration of, 70, 296–97*t*
  orbits, 115–16, 117
  physical properties, 27, 28*t*, 29
  spacing from Sun, 35–36, 124*t*
  travel times to, 124*t*
plants:
  cultivation of, on starships, 256, 258, 278–79
  effects of space travel on, 47, 48–49
  for planetary atmosphere transformation, 148, 149
plasma physics, 301, 303
Pleiades, 175
Pletesk space center (Soviet Union), 319
Pliny the Elder, 2
Pluto, 4, 8, 23, 27, 36, 117, 176, 220, 253, 261, 263
  exploration of, 157
  manned station on, 274
  physical properties, 28*t*, 32, 172
  travel time to, 122, 123, 124*t*
polarimetric radial velocity meters, 245
polarimetric wavelength calibration, 238
poles:
  celestial, 111 and *illus.*, 112, 213
  ecliptic, 113, 114 *illus.*
  galactic, 211 and *illus.*, 212 and *illus.*
  terrestrial, 108–9 and *illus.*
pollution monitoring, by satellite, 78, 81, 84
POLO (Polar Orbiting Lunar Observatory), 314
Ponnamperuma, Cyril, 14
Popma, Dan, 50, 52
population, increases in Earth's, 81, 83
Portuguese explorations, 2
precession, 112
pressurized-water reactors (PWR), 304
Project Cyclops, 20–24, 183, 254
Project Daedalus, 196–98, 204 *illus.*, 250, 263
Project Orion, 185, 188–91, 194, 197, 204 *illus.*, 284
Project Ozma, 22
projection, of spherical surfaces, 110
proper motions, of stars, 221, 222 *illus.*, 222–23, 223 *illus.*
  Doppler effects, 224, 225 *illus.*
propulsion systems. *See* rocket engines; nuclear rocket engines
protoplanets, 35–36
protostars, 234
protostellar nebulae, 34, 235
proto-Sun, 34–35
Proxima Centauri, 55, 217, 286
pulsars, 270
pulsed fusion rocket engines, 186, 197, 284
Purcell, Edward, 181, 183, 184
pyrolitic-release experiment (Viking), 132–33

quasars, 78, 244

radar beacons, in space, 125

radar detection, of alien starships, 284
radial coordinates, 107
radial velocity, of stars, 221–22, 223, 224
  and extrasolar planets, detection of, 238–39
radiation:
  from nuclear engines, 65, 71 *illus.*
  in space, 40–42, 47. *See also* solar flares.
  and Solar System navigation, 116
radioactive waste disposal, 85–86, 304, 305
radioastronomy, 83, 159
  and the search for life, 13, 16, 17–24, 183
  frequency ranges, 18–20
radioisotopes:
  in gas-fuel reactors, 304
  mapping, dating, of extrasolar planets, 281–82
radio telescopes, 21, 22 and *illus.*, 23 and *illus.*, 24, 159
Ragsdale, Robert G., 140
Ragusa (Italy), 249
RAIR (Ram-Augmented Interstellar Ramjet), 250–51
Reactor-in-Flight Test (RIFT), 67
rectangular coordinates:
  galactic, 212 and *illus.*
  terrestrial, 105–6, 106 *illus.*
rectennas, 80, 82
red blood cell loss, during space travel, 44, 45, 48
red giants, 34, 227–28
Redding, J. L., 192
red shift, 224. *See also* Doppler effects
reference positions, in astronomy, 213
regolith (lunar soil), 95, 96, 101
relativity effects, in interstellar navigation, 213, 226–28
rems, 42
Rhea, 171
right ascension, 111 and *illus.*, 112
rocket engines, 56–73, 312–13
  advanced shuttles, 62–65
  exhaust velocity, 56, 57, 58, 65
  expansion ratio, 60
  fuels, 55–56, 58, 63, 65
  for Mars voyages, 135–36 and *illus.*, 137 *illus.*, 139–40
  mass ratio, 57–58
  payloads, 60–62
  staging, 57, 62
  thrust, 57
  specific impulse, 58, 124, 145, 294, 319
  See also Space Shuttle program; nuclear rocket engines; starships
Rosenblatt, Frank, 236
Ross 614, 230
rotation:
  period of, 108
  poles of, 108, 112
Rover Project (1955), 65–70, 73, 140, 292
Runnels, D. D., 12
Ruppe, Harry O., 64

Sabre Foundation, 318

Sagan, Carl, 17, 162, 184, 234, 235
Sagittarius, 174, 212
Salkeld, Robert J., 63
Salyut-6 space station, 292, 310
Saturn, 8, 27, 99, 110, 123, 169, 170 *illus.*, 171, 295
　moons, 157, 170–71
　physical properties, 28*t*, 31, 32
　ring system, 32
　travel time to, 124*t*
Saturn rockets, 61, 185
　for Mars voyages, 134, 135, 138, 139
Schiaparelli, Giovanni, 8, 128
Schmidt telescope (Palomar Observatory), 31
Schwenk, Francis C., 294
science fiction, 5
Serkowski, K., 238–39, 239 *illus.*, 240, 244
sextants, for interstellar navigation, 227
SETI (Search for Extraterrestrial Intelligence), 10, 18–24
　radio antennae, 22 *illus.*, 23 *illus.*
70 Ophiuchi, 196
silicon, silicon compounds, 7
Sirius, 20, 99, 217, 299, 300
Skylab mission, 42, 43, 44–45, 88, 140, 268, 310
solar day, 158
solar flares, 38, 41–42, 125, 157
solar nebula, 34–35, 164, 165, 194
Solar Neighborhood Institute, 245, 306–7
solar neutrino detector, 156
Solar Optical Telescope, 314
solar-power satellites, 78, 80–82, 318
　specifications, 82, 84
Solar System, 3, 26–36
　comets, 33
　evolution of, 34–36, 78
　exploration:
　　manned, 78, 294–98, 302*t*
　　unmanned, 294–95, 296–97*t*, 302*t*
　inner and outer, 4, 27
　moons, 32–33
　origin, 33–34
　physical properties, 27, 28*t*
　planets, 27–33. *See also* individual planets
　Sun, 26–27. *See also* main entry
Solar System navigation, 105–20. *See also* interplanetary voyages
　coordinate systems, 105–8
　　celestial, 110–12
　　ecliptic, 112–13, 114 *illus.*
　　planetary, 108–10
　distance measuring:
　　laser ranging, 115
　　triangulation, 113–14, 114 *illus.*
　trajectories, 115–20
solar wind, 27, 36, 41, 116, 156, 157, 159, 176, 194
　as source of hydrogen, 96–97
solid-core nuclear engines, 66–67, 70, 124, 140, 302*t*, 303
Solinus, 2

South Pole. *See* poles
Soviet Union, 188, 189, 249, 303
　nuclear power plants, 84
　space programs, 292, 301, 310, 311
　solar-power satellites, 82
space, 3–4, 37
　gas in, 29. *See also* interstellar space
　survival in. *See* weightlessness; life support systems; space medicine
　temperature of, 38–39
Space Age, 2–3, 4
space agencies, 310, 312, 315–17, 320. *See also* NASA; European Space Agency
space antennae, 269–70
space ark, 185–87
space colonies, 79–85
　cost-effectiveness, 80–81, 84–85
　specifications, 80
　transportation, 80
space communication systems, 92–94
space exploration, 5, 123, 127, 155. *See also* Solar System exploration; extrasolar planets
space industries, manufacturing, 88, 91, 95–97, 315, 318
Spacelab, 45, 46 and *illus.*, 47, 312, 314
space medicine, 45–49, 88, 264–65, 267, 276. *See also* weightlessness; life support systems
Space Nuclear Propulsion Office (NASA-AEC), 66, 70
space observatories, 88, 156–57, 159
spaceports, 88, 317, 318–19
space radiation, 40–42, 47
space satellites, 156–57, 311–12, 315–16
Space Shuttle, 45, 46 and *illus.*, 47, 59–62, 75, 76, 97–98, 121, 126, 243, 266, 268, 290, 291, 295, 303–4, 310, 312, 313, 319
　cost-effectiveness, 80, 87
　and Mars expeditions, 139
　military uses, 311
　Orbiter vehicle, 59
　payload costs, 61*t*, 61–62
　rocket engines, 59–62, 63, 69
　satellite maintenance, 88
space stations, 87–91, 159, 292, 310, 311
　construction, 88, 89 and *illus.*, 90
　crew rotation, 90–91
　for Mars expeditions, 139
　purposes, 87–88
　power sources, 91
　radiation shields, 41, 42
Space Task Group, 134–35
Space Telescope (NASA), 219–20, 221, 312
space telescopes, 314
　diffraction limited, 219
　for extrasolar planet detection, 241–44, 272–73
　on starships, 252, 257
Space Transportation System, 61, 76, 80, 94, 139, 290, 293
spaceships, 121–26, 302*t*. *See also* starships.
　design, specifications, 125–26

(spaceships *cont.*)
　lunar vs. Solar System, 125
　for Mars voyages, 135
specific impulse, of rocket engine thrust, 58, 124, 145, 294, 319
speckle interferometry, 219, 237–38, 241, 244, 245
Spencer, Dwain F., 195
spherical coordinates, 106–7, 107 *illus.*, 108–10
spinning interferometer, 243
Spinning Solid Stage rockets, 62
Spitzer, Lyman, 241, 242
Sputnik, 188
SSME (Space Shuttle Main Engine), 56, 59–60, 60*t*, 63
SSTO (Single-Stage-to-Orbit) vehicle, 63–64, 139, 319
staging, of rocket engines, 57, 62
standardization, in spacecraft, 295
Star Trek, 5
stars, 177–79, 306–7. *See also* binary stars; multiple star systems; interstellar navigation
　formation, 9–11, 229, 234
　life cycle, 177
　mass requirements, 229–30
　as navigation beacons, 227–28
　proper motions of, 221, 222 *illus.*, 222–23, 223 *illus.*
　types, 177, 178*t*
starships, 188–203, 204–5 *illus.*, 206–7, 249–60
　Animal Colony, 256, 277–78
　communication with Solar System, 253–54, 269–70
　construction:
　　materials, 252
　　sites, 262–63
　costs, 270–71
　crew, 263–67, 275–77
　　age requirements, 264
　　families, 265
　　rotation of duties, 267
　　size, 265–66
　　specialties, skills, 266–67
　　supervisors, 267
　early projects, 188–91
　educational facilities, 264, 268
　hospitals, medical care, 259, 264–65, 267, 276
　information center, 257–58
　life-support systems, 254–56, 266
　maintenance, 256
　observatory, 252, 257, 268
　operation, 268–77
　propulsion systems, 180–207, 250–53
　　annihilation-powered, 203, 205–7
　　catalytic nuclear ramjets, 201–3, 204 *illus.*, 251–53
　　fusion-powered, 184–85, 186, 194–96
　　gravity-assisted, 193–94
　　interstellar ramjets, 186, 197, 198, 250–51, 301, 302*t*
　　laser-powered, 191–93
　recreation center, 258–59
　research on board, 257–58, 259–60, 279
　waste disposal, 256

Steigman, Gary, 205, 206
Stellar Data Center (Strasbourg), 307
stellar flares, 236. *See also* solar flares
stellar parallax, 214–15, 216, 217 *illus.*, 218, 219
stress, in interstellar travel, 188
Stuhlinger, Eric, 192–93
Stull, M. A., 243
Suffolk, Graham C. J., 232
Sun, 26–27, 155–57. *See also* Solar System; solar wind; solar flares
　chemical composition, 26–27
　chromosphere, 27
　corona, 27
　evolution, 34–35
　fusion reaction, 27, 34, 156
　gravitational force, 116
　location, in Milky Way Galaxy, 175
　monitoring of, 155–57, 306–7
　motion of, and radial velocity differences, 238
　photosphere, 26
　relative motion of, 220, 224
　relative speed, 221
　type of star, 177, 178*t*
Syncom IV satellite, 61–62
Syrtis Major (Mars), 128

tangential velocity, of stars, 222–23 and *illus.*, 224
Tau Ceti, 18, 20, 22, 239
Taylor, Theodore, 188–90
TD-1 satellite, 312
telescopes, 128
　celestial setting of, 112
　detection of alien starships, 284
　next-generation, 244, 245 and *illus.*, 306–7
　See also radiotelescopes; space telescopes
television broadcasting, by satellite, 92–93, 314
Teller, Edward, 5
Tempel II comet, 314
terraforming. *See* planetary engineering
Thaw refractor, 239–40, 244
thorium, in gas-fuel reactors, 305
thrust, in rocket engines, 57, 58, 193
time dilation, in interstellar travel, 228, 264
Titan, 33, 170 *illus.*, 171, 295
　atmosphere, 171
　as exploration base, 157
Titania, 170, 171
　as exploration base, 157
Titan III-E rockets, 58, 62
torus, 52–53
trace elements, and the origin of life, 12–13
Trans-World Airlines, 61
Trapped Radiation Belt (Jupiter), 167
triangulation, 113–14, 114 *illus.*
Triton, 33, 157, 171, 172
Trojan asteroids, 163
Tyuratam space center (Soviet Union), 319

UFOs, 283–84

Ulam, Stanley, 188, 190
Ulrych, Tadeusz, 232
ultraplanetary probe, 194
UMDH (unsymmetrical dimethyl hydrazine), 313
United Aircraft Research Laboratories, 71–72
United States:
   energy consumption, 80, 82, 85
   space programs, 69–70, 290–91, 294. See also NASA.
     budget, 317
Universe, mapping of, 220, 268, 279
Uranus, 8, 27, 167, 295
   moons, 157
   physical properties, 28t, 31, 32

Valle Marineris (Mars), 142, 144 illus.
Van Allen, Dr. James, 40–41
Van Allen Belts, 40–41, 87, 167
van de Kamp, Peter, 230, 231, 232
Vandenburg Space Center, 311
Vanguard I, 58
van Maanen's Star, 239
Vega, 224
velocity, determination of, in interstellar navigation, 227
Venus, 23, 27, 157, 160–62, 161 illus.
   atmosphere, 8, 30–31, 160, 162
   manned voyages to, 70, 123, 133, 134
     von Braun plan, 137–41
   navigation to, 117, 118 illus.
   physical properties, 28t, 30–31
   rotation period, 160, 162
vernal equinox, 111 illus., 112, 113
Verne, Jules, 5
Vesta, 165
vestibular disturbances, during space travel, 44
Viewing, D. R. J., 283
Viking spacecraft, 8, 55, 57, 131–33, 139, 140, 142, 143 illus., 144, 145 illus., 150, 285
   biology experiments on Mars, 132–33
Virgo Supercluster, of galaxies, 4, 179, 220
viruses, 7, 14–15
VLBI (very long baseline interferometry), 236–37, 270
Voltaire, 26

von Braun, Werner, planetary expedition plan, 135–41
von Hoerner, Sebastian, 181–82, 184, 185, 190
Voskhod mission, 43
Vostok mission, 43
Voyager missions, to Jupiter, 32, 55, 166 illus., 168 illus., 194, 321

Ward, William, 234
water:
   on asteroids, 164
   on earth, 29
   on Jupiter's moons, 167
   life based on, 7–8
   on Mars, 142, 143 illus., 144, 151
   on Saturn's moons, 171
Watson, Fletcher, 39
weightlessness, 42–44, 78, 98, 274–75
   animal experiments, 47–48
   partial g effect, 47, 52–53
   physical effects, 32–44
   recovery times, 44
   in space stations, 87
weight loss, during space travel, 44
Wells, H. G., 5
Wertz, James R., 227–28
Western Europe, space programs, 311–12, 317. See also European Space Agency
Westinghouse reactors, 67, 68–69
Whipple, Fred L., 39, 40
Whitmire, Daniel P., 202, 251, 300
wing batteries, for starships, 200–1
Wood, L., 195
World Solar Observatory, 156–57
Wrigley, Walter, 210, 226
wrist radiotelephones, 92

xenon, 132
X rays, 41, 42

Yuri satellite (Japan), 316

zero g. See weightlessness

Adleman
Bound for

DATE DUE

3-29-04

DEMCO, INC. 38-2931

95-1331

TL791 .A33

Adelman, Saul J.
Bound for the stars.